THE 350-FOOT-HIGH LAVA BLUFFS OF PALISADE HEAD, LAKE SUPERIOR

ASPENS AND PAPER BIRCHES IN FALL FOLIAGE

REEDS AND WATER LILIES ON SNOWBANK LAKE

MOSS, FERNS, BUNCHBERRY AND ASTERS NEAR GUNFLINT TRAIL

WHITE SPRUCES ATOP A HARD-PACKED SNOW RIDGE, MANITOBA

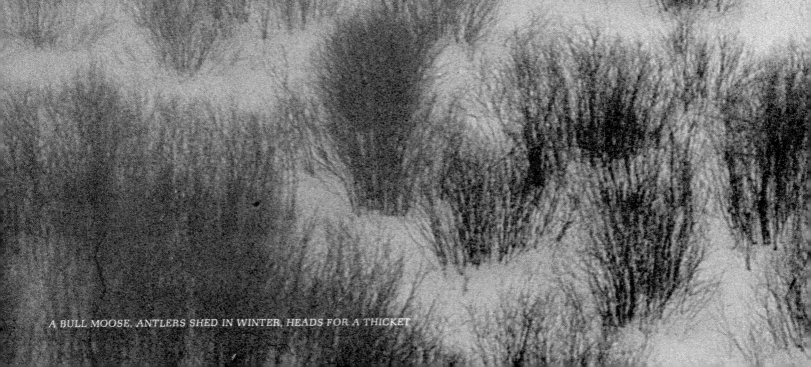

A BULL MOOSE, ANTLERS SHED IN WINTER, HEADS FOR A THICKET

A STAND OF PAPER BIRCHES IN SUMMERTIME

RAPIDS ON THE GRANITE RIVER, SEEN FROM A PORTAGE TRAIL

TIME
LIFE
BOOKS ®

LIFE WORLD LIBRARY
LIFE NATURE LIBRARY
TIME READING PROGRAM
THE LIFE HISTORY OF THE UNITED STATES
LIFE SCIENCE LIBRARY
GREAT AGES OF MAN
TIME-LIFE LIBRARY OF ART
TIME-LIFE LIBRARY OF AMERICA
FOODS OF THE WORLD
THIS FABULOUS CENTURY
LIFE LIBRARY OF PHOTOGRAPHY
THE TIME-LIFE ENCYCLOPEDIA OF GARDENING
THE AMERICAN WILDERNESS
THE EMERGENCE OF MAN
THE OLD WEST
THE ART OF SEWING
FAMILY LIBRARY:
    THE TIME-LIFE BOOK OF FAMILY FINANCE
    THE TIME-LIFE FAMILY LEGAL GUIDE

THE NORTH WOODS

THE AMERICAN WILDERNESS/TIME-LIFE BOOKS/NEW YORK

BY PERCY KNAUTH
AND THE EDITORS OF TIME-LIFE BOOKS

TIME-LIFE BOOKS

FOUNDER: Henry R. Luce 1898-1967

Editor-in-Chief: Hedley Donovan
Chairman of the Board: Andrew Heiskell
President: James R. Shepley
Chairman, Executive Committee: James A. Linen
Group Vice President: Rhett Austell

Vice Chairman: Roy E. Larsen

MANAGING EDITOR: Jerry Korn
Assistant Managing Editors: David Maness,
Martin Mann, A. B. C. Whipple
Planning Director: Oliver E. Allen
Art Director: Sheldon Cotler
Chief of Research: Beatrice T. Dobie
Director of Photography: Melvin L. Scott
Senior Text Editors: Diana Hirsh, Ogden Tanner
Assistant Art Director: Arnold C. Holeywell

PUBLISHER: Joan D. Manley
General Manager: John D. McSweeney
Business Manager: John Steven Maxwell
Sales Director: Carl G. Jaeger
Promotion Director: Paul R. Stewart
Public Relations Director: Nicholas Benton

THE AMERICAN WILDERNESS
SERIES EDITOR: Charles Osborne
Editorial Staff for *The North Woods*:
Text Editor: Jay Brennan
Picture Editor: Susan Rayfield
Designer: Charles Mikolaycak
Staff Writer: Gerald Simons
Chief Researcher: Martha T. Goolrick
Researchers: Angela Dews, Terry Drucker,
Barbara Ensrud, Rhea Finkelstein,
Villette Harris, Myra Mangan,
Mollie E. C. Webster
Design Assistant: Mervyn Clay

Editorial Production
Production Editor: Douglas B. Graham
Assistant: Gennaro C. Esposito
Quality Director: Robert L. Young
Assistant: James J. Cox
Copy Staff: Rosalind Stubenberg (chief),
Eleanore W. Karsten, Florence Keith
Picture Department: Dolores A. Littles,
Joan Lynch

Valuable assistance was given by the following
departments and individuals of Time Inc.:
Editorial Production, Norman Airey,
Nicholas Costino Jr.; Library, Benjamin Lightman;
Picture Collection, Doris O'Neil;
Photographic Laboratory, George Karas.

The Author: Percy Knauth, a freelance writer and editor, has had a special affection for the north woods since he first sampled the region as a boy. He has returned many times, and he reexplored several out-of-the-way areas for this book. He has been a writer and editor for TIME, SPORTS ILLUSTRATED and LIFE, and served for three years as associate editor of the LIFE Nature Library.

The Consultant: Sigurd F. Olson is widely regarded as the dean of north woods naturalists. A former professor of biology and ecology, he has for many years devoted himself to the study and preservation of wilderness. He has been a long-time consultant to the Secretary of the Interior and the National Park Service, is a former president of the Wilderness Society and author of seven books, chiefly about the north woods. He lives in Ely, Minnesota, close to the forests of the Quetico-Superior lake country.

The Cover: A late spring dusk settles on a rain-soaked forest in northeastern Minnesota. This dense woodland along the Baptism River near the north shore of Lake Superior contains several species of trees found through the north woods. In the foreground paper birches, their trunks bent by the weight of winter snow, set off the two sturdier forms in the middle background, white spruce *(center left)* and white pine *(center right)*. The smaller evergreens in the foreground are balsam firs.

Contents

A Land of Woods and Water

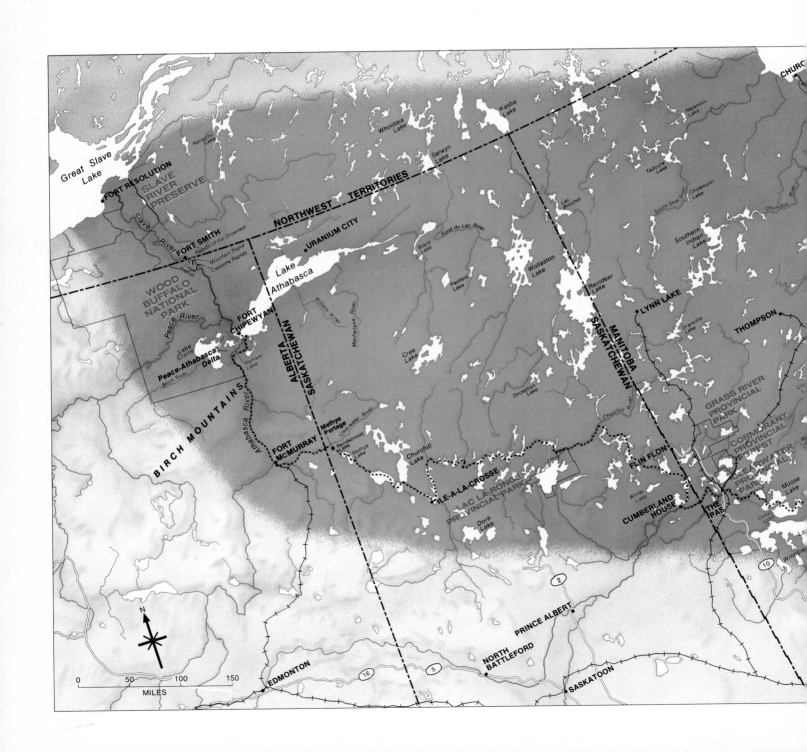

Lying about midway in the great coniferous forest belt that spans the North American continent from Labrador to Alaska, the north woods described in this book extend from northern Minnesota to embrace parts of four Canadian provinces as well as the Northwest Territories (green shading, right). As evident in the detailed map below, lakes and rivers abound in the region, giving it a look as much of waterscape as of forested landscape. A combined water-and-land route, some 2,000 miles long, helped foster the celebrated fur trade of the 18th and 19th Centuries; this historic Voyageurs' Highway, as it is known, is shown by a dotted line between its terminal points, Grand Portage on Lake Superior and Fort Chipewyan on Lake Athabasca in the northwest. Outlined in red and marked with red type are national parks, provincial parks and other protected wilderness areas.

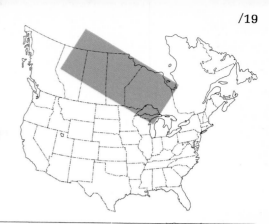

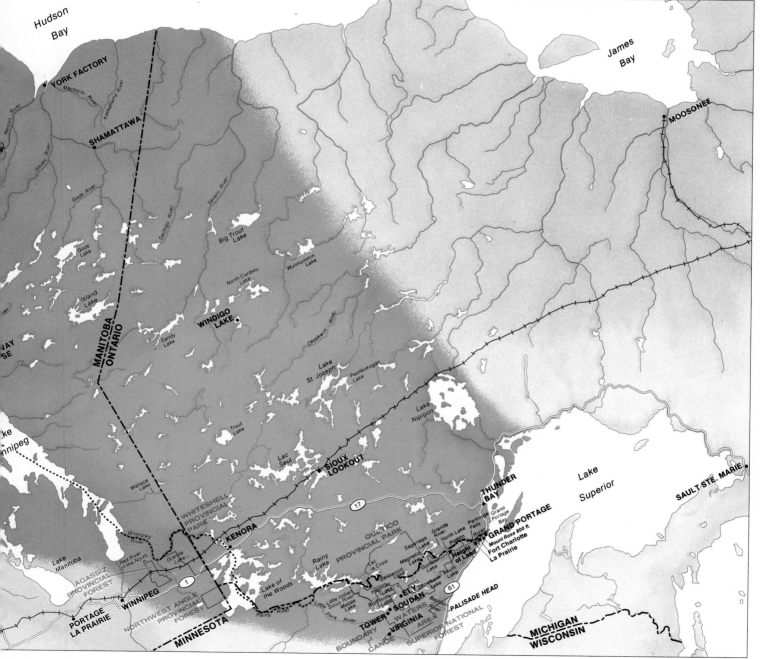

1/ An Infinity of Trees and Water

*In the woods a man casts off
his years, as the snake his slough....*

RALPH WALDO EMERSON/ NATURE

Every man, when he surveys the places he has been in his life and reflects on the experiences he has had there, will choose one place above
all others where he likes to be. Such a place, for me, is the north woods.
This vast and marvelous land of forest, lake and stream has no precise
boundaries, nor does it need any: it exists as much in men's minds as it
does on the North American continent. My favorite north woods—with
which this book will be primarily concerned—lie at just about the center of the great forest belt that stretches from Labrador to Alaska. The
part of it that most appeals to me begins on the northern fringes of the
Quetico-Superior area, straddling the international border between
Minnesota and the Canadian province of Ontario. From here it extends
about 800 miles northwest to the western shore of Hudson Bay, from
there west about 600 miles across the provinces of Manitoba and Saskatchewan to Lake Athabasca, and from there about 1,100 crow-flight
miles southeastward back to the Minnesota border. The shape of this
domain is very roughly that of an enormous triangle, with one of the
angles, fittingly, at the historic settlement of Grand Portage.

Where Grand Portage lies, at the edge of the great forest in northern
Minnesota, Lake Superior beats with icy fingers against a rocky shore.
A small creek flows out of the hills here, meandering past cultivated
fields through groves of willow and aspen. Bearberries, blueberries and
wild roses crowd in on the beach; beyond them, like clouds, soar the

tops of sentinel pines. In a meadow by the shore, weathered and gray, stands the stockade of an old fort. These high spiked pickets and the buildings they enclose—reconstructed in modern times—commemorate a period, beginning more than 200 years ago, when the trail that starts here was the jumping-off place for one of the epic adventures in North American history. It was from here that the incredibly hardy breed of French Canadians known as the voyageurs challenged the most forbidding wilderness on the continent, traveling on foot and in birch-bark canoes, repeatedly making a 2,000-mile journey into the north woods in search of the furs that were to make other men's fortunes.

A small hill stands behind the stockade at Grand Portage, an 800-foot eminence known today as Mount Rose. There is a path that leads up it, wandering back and forth for perhaps half a mile as it climbs toward the summit, where in the old days lookouts were posted to shout the arrival of the canoe brigades. From the top, looking southward, you see the seemingly endless expanse of Lake Superior. Toward the north, however, the view encompasses a haunting panorama: range upon range of dark-green forested hills, all of them bearing that ineffable stamp of the wilderness, an aura of solitude and of perfect peace.

When North America was first being explored by white men, the north woods stretched across the continent, forming an immense dark roof of spruces and pines. Where the climax forest stood, trees as straight and stately as the columns of a Gothic cathedral rose sometimes 100 feet and more; about halfway up they branched out to form a canopy that the sun could penetrate only when it was low on the horizon. Below, the forest floor was carpeted with a thick, springy covering of brown needles that sifted down without cease from the branches high above; with no underbrush to impede the view, long dim vistas unrolled through the trees. Elsewhere, along streams and ponds and in the wetter areas of the forest, grew stands of such deciduous trees as the willow, aspen and, most important to history, the paper birch, *Betula papyrifera*. It was from this ghostly white tree, glimmering through the forest shadows, that the Indians—then the only inhabitants of the wilderness—fashioned their birch-bark canoes.

The north woods country is, indeed, the natural habitat of the canoe. No one knows when the first such craft was built or how it evolved, but the process was surely as inexorable as the one that, long ago and far away in a very different environment, brought forth the wheel. A boat is needed in a land of lakes and rivers and, since all the waterways do not always connect with one another, it must be a boat that can easily be carried overland—portaged. It must also be a boat that can cope with widely differing kinds of water, from the turbulent close quarters of rapids to the long unprotected stretches of large lakes where storms come suddenly, without warning, and whip a glassy surface into white-capped waves. Such a boat is the canoe, with its high bow and stern that smoothly divide onrushing waters, its broad midsection to carry heavy loads, and its light construction that enables its crew to take it on their shoulders and carry it when the waterway is blocked.

The canoe is one symbol of the north woods, and there is another: the beaver. This busy rodent is hardly the most imposing animal in a wilderness where black bears roam, where moose can be seen browsing in the shallows by day and the wolf's howl can be heard at night. But the beaver, more than any other creature, affected the history of this rugged land and indeed sometimes helped make it less difficult to traverse.

Many animals are capable of modifying their environment to suit their particular needs, but beavers are among the most successful at it. With the brushwood from the trees felled by their powerful incisor teeth, they build dams that back up streams, creating ponds where there may have been rapids before, eliminating cascades and other watery hindrances to travel. By raising the levels of ponds and rivers they can make an impassable water route passable. Far in the north, in Manitoba, there is a striking example of this in a shallow river that beavers made navigable across a height of land. Paddlers coming up that stream find themselves crossing one beaver dam after another, until finally they cross one more and discover that the river has now reversed its direction and they are paddling downstream. Long ago the Indians called this river the Echimamish, meaning the-river-that-flows-two-ways, and the name remains. But to keep the passage open, human hands eventually had to take up the beavers' work, because for a period of years the beavers disappeared from the area, their dams deteriorated, the water level dropped and the route became unusable.

What caused the beavers to disappear—not only from the Echimamish, but from all over the north woods—was not any natural catastrophe but their wholesale extermination by man. From a population that once certainly numbered in the millions, the beavers were reduced by trapping almost to the point of extinction. They became exceedingly scarce in Minnesota, in southern Canada and, as their pursuit was intensified, in the more northerly reaches of the land.

The beavers are now coming back, though in nothing like their original numbers; the sight of their dams and ponds is now a familiar one in the forests north of Lake Superior. For today they are protected against excessive trapping by law, secure against a repetition of their earlier fate. That fate was the direct outcome of a single fact about the beaver: the desirability of its pelt. When men first came from afar to seek it, the wilderness of the north was probably the world's richest source of other fur-bearing animals as well—mink, ermine, marten, muskrat, lynx, wolverine, bear. These animals, too, were sought and caught, but the beaver, though its fur was less luxuriant than some of the others, was prized above all. The value of its pelt lay in its use not as fur but as the material from which felt was made—felt that was then molded and pressed into manifold forms of felt hats. The attribute that commended the beaver pelt was the woolly underfur right below the outer layer. Each hair of this underfur was finely barbed, making the pelt tightly woven—a quality that was deemed uniquely suited for the making of thick, firm felt hats to grace the heads of the rich and the fashionable, men and women alike, all across Europe.

And so it was the glittering prospect of a bonanza in furs that led to the opening up of the north woods. Some of those who took part were primarily interested in pure exploration, and specifically in finding the fabled Northwest Passage to the Pacific. But eventually they were outnumbered by the fur traders. To the Frenchmen, Englishmen and Scotsmen who built up this trade, the sometimes deadly dangers it entailed were offset by the fortunes it promised: the value of a single year's fur cargo shipped back home could exceed a million dollars.

But the men who bore the heaviest burden of danger, who wrestled with the wilderness in all its unpredictability, did not stand to profit from the fur trade. These men were hired hands, toiling for a wage. It was they who worked the boats, who unloaded the cargoes at the start of a portage and reloaded them at the end, who between these points carried the cargoes on their backs, and who all the while coped with the endless exigencies that nature presents in uncharted terrain—a treacherous current that could capsize a boat, a stretch of swamp that could unbalance a man and his load, a path that proved a dead end and had to be laboriously retraced.

Of these men the most colorful, the most enduring in memory, were the French-Canadian voyageurs. They were in the employ of the North West Company, whose headquarters were in Montreal and which was eventually taken over by the Hudson's Bay Company, an English enterprise. The rivalry between the companies, before the merger in 1821,

Frothing white waters race along the shallow stream bed of the boulder-strewn Little Fork River in northwestern Minnesota.

was intense and often bitter. The Hudson's Bay Company venture into the fur trade started with a number of trading posts on the southwestern shore of Hudson Bay. At first no traveling was necessary; Indians wishing to exchange fur pelts for the assorted wares the company had to offer came to the trading posts. But as intensive trapping reduced the yields in the areas close to the posts the English, like the French before them, had to penetrate into the wilderness to get their furs. Working westward from Hudson Bay, they used boatmen from their own country —stalwart Orkneymen from the islands off Scotland's northern coast. They also used a type of boat different from the canoe—a double-prowed wooden rowboat, 28 to 40 feet long, with a wide, flat bottom. On the bigger rivers and lakes of the north country this craft—called the York boat, after York Factory, the Hudson's Bay trading post where it was built—proved to be workable and durable.

But the York boat, like the men who propelled it, lacked the romantic flair of the canoe and its crew of voyageurs. Cursing and singing and laughing in irrepressible Gallic good humor, the voyageurs made a special mark on the history of the continent. Their traces are to be detected even now in the north woods in the campsites they used and in the portages over which they carried their canoes and cargoes.

The wilderness that the voyageurs knew is much diminished today, the roof that once stretched across the continent pierced and decimated by the white man's civilization. In ecological terms it continues to exist as an entity: it is the topmost of three great forest belts that girdle the North American land mass, its mark of distinction the conifer. Experts in such matters call this coniferous belt the boreal forest—in honor of Boreas, the mythological Greek god of the north wind—or the taiga, after the Russians' word for the subarctic coniferous belt in Siberia and northern Europe. But whether it is called taiga or boreal forest or simply the north woods, the heavy hand of man has changed it. The push of population inland from the coasts, the onslaughts of loggers and miners, the inroads of power-plant builders, have sheared away vast sections of it. In areas where once only canoes could pass there are now shipping routes and railroads and busy multilaned highways.

But some of the great forest remains intact. From Grand Portage up the Pigeon River, past Gunflint and Saganaga and Rainy Lakes, down the Rainy River and into Lake of the Woods, and from here farther northward past Lake Winnipeg—much of this is wild land still. The part most zealously guarded against the pressures of civilization lies within the 16,000-square-mile Quetico-Superior country. Within Quetico-Superior, the 873,847 acres of the Boundary Waters Canoe Area on the U.S. side and the 1,144,960 acres of Quetico Provincial Park on the Canadian side have been set apart by the respective governments as wilderness. Beyond Quetico-Superior, to the north and west, are other immense tracts unprotected by law but largely unexploited.

Viewed from ground level, even a small segment of the north woods looms as an infinity of trees—somber, brooding, monotonous yet overpowering in their mass, a formidable citadel that appears all but impossible to breach. Within the woods one meets still further obstacles not readily seen from outside. Not until men learned to fly could the diversity and difficulties of the interior be fully perceived. From the air it is quickly apparent, for example, that much of the north woods is rock—mostly granitic outcroppings of the Canadian Shield, a portion of the earth's crust so old it dates back to the Precambrian Era, which ended about 600 million years ago. The rock takes forms often perilous to the traveler: sudden steep precipices, massive boulders, sharp ledges that jut out from the banks of a stream or lie concealed beneath it.

Water is a force even more to be reckoned with in this wilderness. It is there in such abundance and variety—gently flowing, tumbling, cascading—that from a plane the north woods appear to be almost as much a waterscape as a forested landscape. The profusion of lakes alone stuns the mind; it is as if some giant brush had scattered droplets everywhere in careless abandon, again and again. Minnesota on its license plates claims 10,000 lakes, Manitoba on its license plates 100,000, and neither seems exaggerated. A few more formal statistics offer a hint —but a mere hint—of the number of lakes in the north woods. Within the 873,847-acre Boundary Waters Canoe Area alone there are some 2,500 lakes that are 10 acres or more in extent. Across the border, as one flies north, the same astonishing profusion of lakes unfolds.

It is difficult to see how anyone not equipped with detailed maps and accurate navigational instruments could ever have found a path through this labyrinth of water. The early voyageurs, of course, had only crude maps, if any, and little in the way of instruments beyond a compass. How, then, did they find their way from lake to lake? How could they have known where lay the portage that would carry them onward to the next leg of their waterway? How did they know which river flowed northwestward? Or which southeastward? Or which river was passable, and which held rapids that could not be run? How did they find a pattern where seemingly no pattern exists, as one looks down upon this staggering plenitude of sparkling water?

Today, with the help of modern maps based on aerial surveys, the canoeist is prepared for some of the problems the voyageurs faced. One complexity with which they had to deal concerned currents. As they proceeded northwestward from Grand Portage, they were paddling upstream against currents that flowed east or east by south; then for a while they had generally favorable currents; but in the second half of their journey, after they had passed Lake Winnipeg and were approaching the region of Lake Athabasca, they found themselves again fighting upstream on rivers and streams that flowed east or east by south.

The voyageurs found the pattern of the waterways by trial and error and by another equally effective means: they tapped the knowledge of the country possessed by the Indians. The French were the first white men to colonize Canada; by the time the fur trade began to develop, many of them were natives of the country—French Canadians whose homes were in peaceful villages along the Saint Lawrence Valley. An easygoing people, they had a talent for getting along with the Indians. They made friends with them, intermarried with them and in general treated them as brothers. Thus they were able to benefit from the experience of generations of hunting Indians who had long since developed their own canoe routes and portages. Most of the trails followed by canoeists today are portage paths trampled smooth by moccasined Indian feet centuries before the first white men arrived.

One vital lesson of the wilderness that the voyageurs had to learn at first hand concerned the climate of these latitudes. However familiar they might become with waterway patterns, the knowledge was futile unless it could be acted upon within the five months—roughly May 1 to October 1—between spring thaw and autumn freeze; the rest of the year their route was blocked by ice. Moreover, the time of thaw or freeze was never certain, and winter's premature arrival could catch a canoe brigade far from its base, forced to try to survive in some hastily improvised forest shelter.

To deal with this problem, the officials of the Montreal fur trade divided the 3,000-mile route from Montreal to Lake Athabasca into two stages, each traveled by a separate set of voyageurs. Some of the canoe brigades paddled from Montreal to Grand Portage, carrying trade goods, and from there back to Montreal, carrying furs. The fur cargoes were brought down to Grand Portage by canoe brigades that were based at Lake Athabasca or other northern outposts, and that would then return to these bases with the trade goods picked up at Grand Portage.

The 1,000-mile journey from Montreal to Grand Portage actually began a few miles upstream at the village of Lachine, at the head of a long stretch of fast water that barred the canoes from getting closer to the city. Here the goods that were to be traded to the Indians were wrapped in canvas in standard 90-pound bales. Included in the cargo were iron stoves, cooking pots, blankets, cloth of all kinds, needles, thread, pins, flour, salt, cheap trinkets, beads, guns, gunpowder, ammunition, brandy, rum, wine—all the blessings an advanced civilization could muster to tempt these Stone and Copper Age people who held the key to huge fortunes in furs. Fully three tons of cargo was loaded into each canoe —the great, 36-foot *canot de maître,* also known as the Montreal canoe, paddled by a crew of eight to 10 men. Whole fleets of these craft, built entirely of birch bark on a light cedar frame, set out in company, paddling up the Ottawa River until it turned northward, then taking the Mattawa River to Trout Lake, then portaging over a height of land to Lake Nipissing, which led the voyageurs down the French River to Georgian Bay in Lake Huron. They paddled through the North Channel of that lake to Sault Ste. Marie, portaged around the rapids there into Lake Superior, and then faced 450 miles of treacherous north-shore water to Grand Portage. The trip took six to eight weeks: departing from Montreal just after the breakup of the ice around May 1, the voyageurs arrived at Grand Portage toward the end of June.

It is hard to believe that anyone could belittle this journey, but the voyageurs who came down from the northlands did so: they held the Montreal brigades in contempt as *mangeurs de lard,* or pork eaters, who spent their winters enjoying the softer life at home. The *hommes du nord,* or men of the north, as they liked to call themselves, spent their winters in the rugged outposts of the fur country, places like Cumberland House on the Saskatchewan River, or Ile-à-la-Crosse on the Churchill River, or Fort Chipewyan, the end of the line on Lake Athabasca. From these points, as soon as the ice broke up, they brought down the furs that had been acquired over the winter. Packed in the standard 90-pound bale, these were transported in the smaller, 25-foot North canoes, each with a crew of five or six men. If the ice broke up on the Athabasca River by May 15, the *hommes du nord* usually made the entire 2,000-mile trip to Grand Portage; if spring was late in coming, they would have to meet at Rainy Lake with a special detachment sent up from Grand Portage to take the furs (continued on page 32)

While mushrooms may be found in any area that supports plant life, including lawns and sand dunes, by far the favorite habitat of these fascinating fungi is a forest floor. Nowhere else do they thrive in such diversity of species, sizes, shapes and colors. The 11 different kinds of mushrooms shown at right and on the following pages are but a fraction of the hundreds of species that grow in the north woods. Some are delicious but deadly, others delicious and edible, but all share a vital peculiarity that explains the forest's attraction for them. Mushrooms lack chlorophyll, and so cannot use sunlight to manufacture their food. Instead they must feed on organic compounds: living trees, dead logs, and decaying leaves and needles —all abounding on the forest floor.

POLYPORUS ALVEOLARIS (HONEY-COMBED POLYPORE)

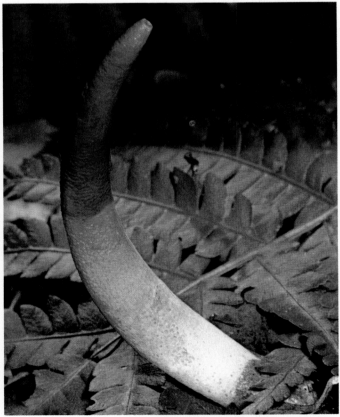

LECCINUM AURANTIACUM (ORANGE BOLETE)

MUTINUS CANINUS (DOG STINKHORN)

POLYSTICTUS TOMENTOSUS (HAIRY POLYPORE)

PHLOGIOTIS HELVELLOIDES (NO COMMON NAME)

SARCOSCYPHA COCCINEA (SCARLET CUP)

MYCENA LEAIANA (LEA'S MYCENA)

HERICIUM CORALLOIDES (CORAL HYDNUM)

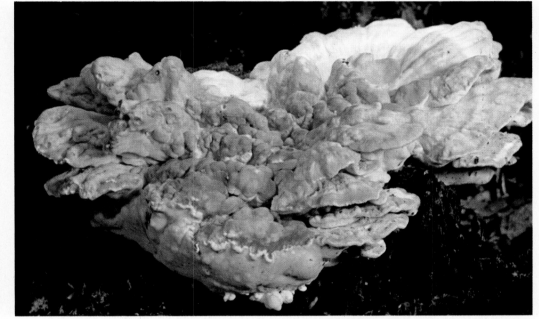

LAETIPORUS SULPHUREUS (SULPHUR SHELF)

CLAVARIA PULCHRA (BEAUTIFUL CORAL)

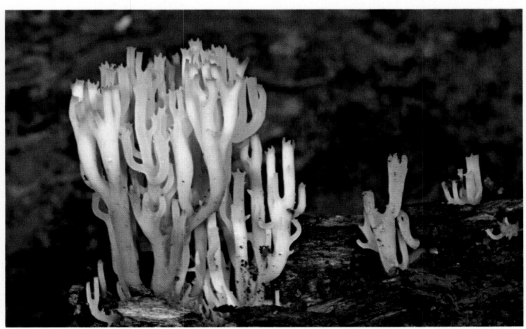

CLAVICORONA PYXIDATA (CUP-BEARING CLAVARIA)

on down, since otherwise they would be unable to get back north before the freeze in October.

Usually, however, the final exchange of cargoes took place at Grand Portage in mid-July. This occasion was called the *rendezvous,* and the little fort on Lake Superior at this time played host to at least 1,000 roistering voyageurs and often twice that many Indians. When the *rendezvous* was over, the much-abused "pork eaters" took the furs back to Montreal in the big canoes, arriving just before the waterways froze and consigned them to the comforts of home. The furs were unloaded, sorted as to quality, and then shipped out to markets all over Europe.

The men of the north, meanwhile, marked the end of the *rendezvous* by setting out on that nine-mile hike up the Grand Portage trail that was the beginning of their journey north.

The Grand Portage trail was a special proving ground for the voyageurs, a place where they tested their strength against each other, where the men were separated from the boys. Each man was required to carry eight bales up to the Pigeon River, the first waterway on the route north; for anything beyond that load he was paid one Spanish dollar per bale. If a man took the normal load of two bales, or 180 pounds, that meant four round trips up and down the trail, a distance of 72 miles. To avoid this the voyageurs sometimes carried more, and sometimes they challenged each other to see who could carry the most. History has preserved, over a century, the name of one George Bonga, who on a wager once carried 820 pounds, or more than nine bales, one mile uphill. It is not likely that anyone ever managed to carry more.

The Grand Portage trail starts out innocuously enough, emerging from a gate in the southeast corner of the fort and following, for a while, the little creek that comes meandering out of the hills. It passes under the shadow of Mount Rose, through thickets of willow and aspen, then through a grove of pine trees and past fields that doubtless were already cultivated in the voyageurs' time. The people whom the hiker may see here today are probably not much different from those who lived in the area in those days, except in dress, for Grand Portage is now part of a reservation for the Chippewa tribe, once a proud race among whom many a voyageur found a bride. Today the Chippewa linger as a remnant, making their living out of souvenirs, or small-time farming, logging or fishing, and once in a while as guides.

Once past the fields, the trail begins to climb steadily toward a notch in the hills ahead that represents a sort of pass. From this height the voyageurs could look back a mile or so to Lake Superior, spreading out all silver in the distance. Here, too, they could take one last look at the bit of civilization represented by the fort below. Then the trail plunges into the forest, not to emerge again for eight miles, when it finally reaches the Pigeon River.

The Pigeon is a paradoxical stream. For most of its length it peacefully wanders past reedy banks over a bed of mud that on occasion is within inches of the bottom of the canoe. At such places, it is full of wildfowl, with ducks clattering up as the canoe rounds a bend, or kingfishers rocketing into its calm surface for their dinner, or a blue heron poking its long bill among the reeds. Mink and muskrat live along the banks; and beavers are there too, feasting on the succulent birches and aspens that grow along the shore. Then, suddenly, the river's mood changes: it goes berserk on rapids, or plunges abruptly over a precipice where its placid brown water turns to roaring yellow foam. Below the point where the Grand Portage trail reaches it, the river becomes completely uncontrollable, erupting in a series of wild cascades through gloomy, echoing canyons where no canoe could possibly survive. Few journals of the voyageur era mention this wild side of the river's character. Perhaps it is because this part of the Pigeon River is like a law of nature: there is no arguing with it, hence no point in discussing it. The Pigeon turns unnavigable; that is all anyone really needs to know.

Where the Grand Portage trail ends there are grass-covered mounds that mark the site of Fort Charlotte, an outpost on the Pigeon River that was the voyageurs' last contact with civilization. Here they beached their canoes when they came down from the north, turning them over against the rain. Here, on the way back north, they would reload the canoes with their cargoes of trade goods. A ton and a half to each 25-foot canoe—that was the standard load for the long journey into the wilds. And each canoe, before it was loaded, was overhauled and any damage from the previous journey carefully repaired.

Fort Charlotte today broods quietly beneath its sheltering pines and spruces, a place far off the beaten track and seemingly forgotten. But there are those who enjoy its history. On any summer weekend, a party of Canadian high-school kids from nearby Thunder Bay may erupt from the woods, having loped up the trail from Lake Superior in the manner of the voyageurs, who never walked when they could run; or a family from the States who have hiked up from Highway 61 may come trudging along and proceed to picnic there. At such times, with people poking animatedly about for relics, it is possible to imagine the fort as a going concern with its high stockade, its log buildings, the gaily painted canoes drawn up on the banks, the lithe voyageurs bending to their final tasks, arranging the 90-pound bales, shouting, cursing, singing snatches of the ballads they would later chorus as they paddled along.

A mile and a half upriver from Fort Charlotte is Partridge Falls, a somber cascade where the Pigeon River, in one of its abrupt changes of mood, leaves its reedy banks and plunges about 50 feet into a ravine. There is a portage through the woods here, perhaps a quarter of a mile long. Alexander Henry the Younger, a North West Company partner who passed this way many times and kept a meticulous journal, mentions it as being "very slippery and muddy," which shows it has not changed much in the 170 or so years since he trudged up it. But another two and a half miles along is a place that every voyageur looked forward to with joy—*La Prairie,* or The Meadow, a wide, grassy place where the canoes could be beached for the night and a *régal* held. Here the men regaled themselves on rum or brandy to celebrate the start of the journey north. "All were merry," Henry noted of one occasion, adding that "there was plenty of elbow room for the men's antics."

From here on it was paddle and portage, paddle and portage, for two or three more days until Height of Land, between North and South Lakes, was reached. Here the upstream struggle ended for a while, for this is the Laurentian Divide, one of the great divides separating the northland's drainage basins. At Height of Land the voyageurs always held a ceremony that was their equivalent of the crossing of the Equator celebrated by salt-water sailors. New men in the brigade were sprinkled with water from a cedar branch dipped in lake water and were sworn to membership in the fraternity, which entitled them to supply drinks for all present. "At this place," noted Henry in his journal, "the men generally finish their small kegs of liquor and fight many a battle."

In a sense, Height of Land marked a jumping-off place: from now on the region through which the voyageurs moved would be subtly different. Rocky points and wooded points, islands innumerable and long vistas of blue water between rolling hills would pass before their prows. By day they might hear the screech of a bald eagle or see an osprey plunging into a lake. By night they would hear the loon's call, that marvelously evocative wilderness sound that at times seems to express unutterable loneliness and brings shivers to the spine, at other times rises to a chorus that blends and mingles the haunting notes much as the northern lights blend and mingle in a midnight sky.

Past Gunflint Lake, past Saganaga, past Basswood and Lac La Croix, the Voyageurs' Highway—as history has named their route—led to Rainy Lake and the Rainy River, and thence to the great Lake of the Woods and down the Winnipeg River to Lake Winnipeg itself. Everywhere there was the same landscape—the tall pines, the graceful spruces, the glimmering white of birches, the trembling silver of aspens. Rocks, cliffs, promontories appeared ahead and disappeared behind. The voyageurs swept down rapids where they could safely run them, and sometimes spilled where they could not. They grunted and swore over the portages. Beaches of sand and of rock received them when at last they stopped paddling at night; they slept exhausted, sprawled beneath their canoes.

From Lake Winnipeg onward, the country into which they pushed was dark with distance, dim with the uncertainty of the unknown and unexplored. Here they encountered rivers—the Saskatchewan and the Churchill—that flowed eastward toward Hudson Bay, across their path as they paddled northwestward. Here, in this wilderness, the moose come down to the water unafraid, and Canada geese trace their course through the skies, their honking sound like a crying wind. The farthest point of the voyageurs' journey was Fort Chipewyan on Lake Athabasca. The north woods stretch even farther northward, beyond Great Slave Lake to Great Bear. But this was beyond the distance that could be accomplished between spring thaw and autumn freeze in a freight canoe. In this far region it is winter more than half the year. What lies beyond is the Arctic, and that is a different story.

The voyageurs have been gone for over 100 years and their songs are heard no more among the tall, dark trees. But the woods in which they spent their lives and that they loved beyond any other place on earth —these are still wild, and so are the lakes and rivers that bore their canoes. In that rugged landscape lies the evidence of times long before the white men came, of times even before the Indians.

Fire in the Woods

When fire rages through deep woods in a storm of incandescence, its roar assaults the ear a mile away. No witness can help but respond with a sense of terror. More practically, men feel a deep responsibility to prevent the outbreak of fire and the destruction it leaves behind.

Yet in recent years many forest ecologists have come to believe that fire can contribute to the health and growth of the natural forest. The subject is a controversial one, however, and proponents admit that they need to know more about the long-term effects of fire. Opponents argue that although fire releases nutrients to the soil, a significant proportion of them is lost in runoff. Plant life may take as long as a century to recover, they say, and many forms of animal life may lose their grazing areas.

Other scientists, however, insist that the basic character of the forest is determined by fire. Certainly some trees such as the aspen, birch and jack pine, though vulnerable to the flames, survive hardily as species through periodic ordeals by fire; other species such as red pines and old white pines have evolved ways of surviving as individuals. For highly vulnerable species such as spruce and fir, healthy growth depends on the virtual absence of forest fire.

In the four north woods areas shown on the following pages—all near the Canadian border in Minnesota—fire has influenced the nature of the forest in ways that bear out the new attitude toward a force long regarded as an unmixed bane. Taken together, these areas show a progression of growth that makes clear the importance of fire.

Even a bad fire can have its beneficent side. A recent example was a serious outbreak east of the Little Indian Sioux River, shown on the opposite page and—in its aftermath—on pages 38 through 41. Driven by a relentless wind gusting up to 30 miles an hour, the Little Sioux blaze raged for three days, destroying nearly 25 square miles of woodland. The conflagration had been, in part, a crown fire—one that consumes the topmost foliage of trees. With their crowns burned off, most of these trees were doomed. Others would die because the fire had cooked their cambium, the layer of growing cells just beneath the bark. As the embers cooled, the scene seemed desolate.

But within months—and in some places after only a few weeks—fresh growth of pioneer species appeared on the site of the Little Sioux blaze.

Propelled by near-gale winds through the treetops, a crown fire consumes a century-old forest of balsam firs, white spruces and paper birches near the Little Indian Sioux River in Minnesota.

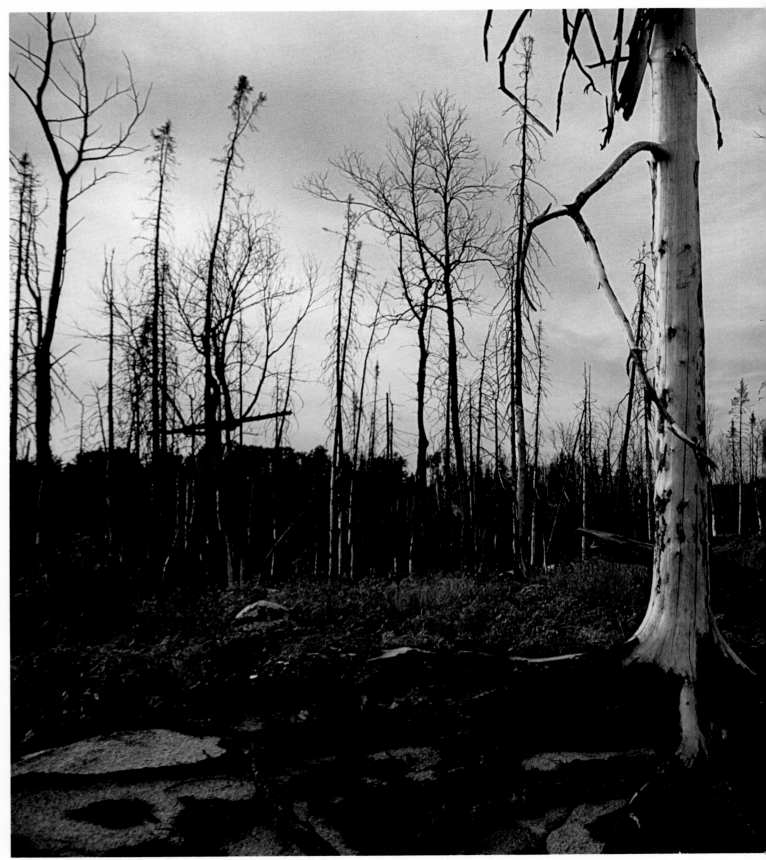

A year after the Little Sioux burn, charred trunks rise like toothpicks from the ashes. Now, however, with the destruction of the forest's

dense canopy, several hours of sunlight reach the forest floor every day, and new growth, visible beneath the dead trees, is proliferating.

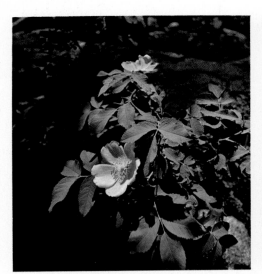

A wild rose flourishes in the fire's aftermath.

Jack-pine seedlings grow next to a log.

Resurgent Growth on the Charred Floor

Within two weeks after the Little Sioux fire had been extinguished in late spring, Forest Service researchers spotted grasses, sedges and yellow bluebead lilies peeking upward through the ashes. Around the end of June, bracken ferns and aspen sprouts began to appear, and jack-pine seedlings were germinating—in some areas at a rate of 20,000 an acre. Wild flowers and shrubs were also abundant. By the end of that summer, the entire area was covered with waist-high vegetation, much of which was young aspen sprouts that had shot up to an average height of four feet since the fire. A year after the holocaust had almost leveled the forest, the photographs on these pages were taken.

This phenomenal growth was a result of several factors. To grow, plants need nutrients like phosphorus, calcium and magnesium. Ordinarily, the bulk of these are locked up in the vegetation—either in the living plants and trees, or in the detritus decomposing on the forest floor. Fire rapidly breaks down these nutrients. Dissolved in rain water, some are lost in runoff. The rest enrich the soil, giving plants a burst of energy. In a brief period, fire can duplicate the work of years of bacterial decomposition. Simultaneously, the obliteration of the dense forest canopy by fire permits intense photosynthetic action to take place, stimulating sun-loving plants to surge from the exposed floor of the forest.

A young ground pine—commonly found in shade—thrives on the Little Sioux burn site.

With its broad leaves opened toward the sun, a new aspen sucker stands against a stark background of fire-blackened tree trunks.

Aspens grow thickly where flames had devastated the woods 20 years before.

A Forest Burn
Two Decades Later

In the 1990s, barring another fire, the site of the Little Sioux blaze will very probably be largely covered with jack pine and trees similar to those shown here. Photographed in Minnesota's Superior National Forest 20 years after an extensive burn, these are dense young stands of aspens and birches, both of which are particularly vulnerable to the ravages of fire. Their thin bark provides little protection during a blaze, and the trees die after their vital inner layer of cambium is subjected to the brief but intense heat of a forest fire.

But over generations of exposure to fire, both species have evolved in such a manner that they are always among the first trees to rise from the ashes. Stimulated by the warmth of the forest floor and encouraged by the new nutrients in the soil, the roots of burned aspens rapidly send out suckers, which become the stems of new trees. Similarly, new birches sprout quickly from the bases of dead parent trees.

Where they have been dominant before a fire, birches and aspens gain such a rapid head start on other trees that, 20 years or so after a burn, they again become the dominant species, proliferating in such dense groves that sun-loving trees like red, white and jack pines have little opportunity for growth. In the shadows cast by the birches and aspens, only a few shade-tolerant species, such as spruce, fir and mountain maple, are able to flourish.

Paper birches share a woodland thicket with mountain maple shrubs resplendent in their autumn colors of orange, yellow and scarlet.

A Competition Arbitrated by Fire

While species like aspen and birch have developed the capability of reproducing themselves after a fire, others have evolved means that permit them to survive a blaze. Chief among these in northern Minnesota are mature red pines and old white pines. Both have thick, insulating bark that protects their cambium from the fire's heat. Moreover, their foliage grows atop tall trunks, remote from the tinder of the forest floor, and immune to practically anything but an intense crown fire.

Even when a stand of pines is incinerated, there is almost always a small number of survivors whose seeds, falling to a floor now burned clear of the thick litter of fallen needles that inhibits the growth of plants, can germinate to begin the building of a new pine forest.

But even when fire does not strike for years, the pine forest is not immortal. Beneath the pines' lofty, sun-shielding canopies, as shown in these pictures taken along the Gunflint Trail, such shade-tolerant species as white spruce and balsam fir find it possible to grow on the forest floor. Eventually, the pines reach old age, die and topple. Over their decomposing timbers will rise the new forest of spruces and firs—unless it is destroyed by fire before the pines die. In that case, the pines may return. Or, suited as they are to growth on burn sites, birches and aspens may overtake the pines and become the dominant species.

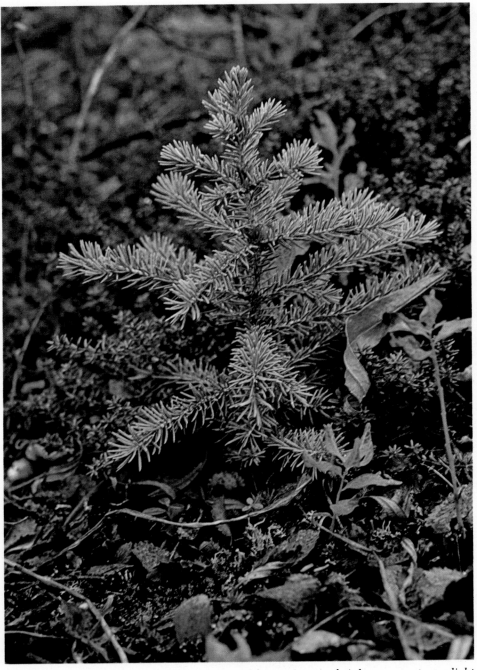

In shadow most of the day, a white-spruce sapling enjoys its brief exposure to sunlight.

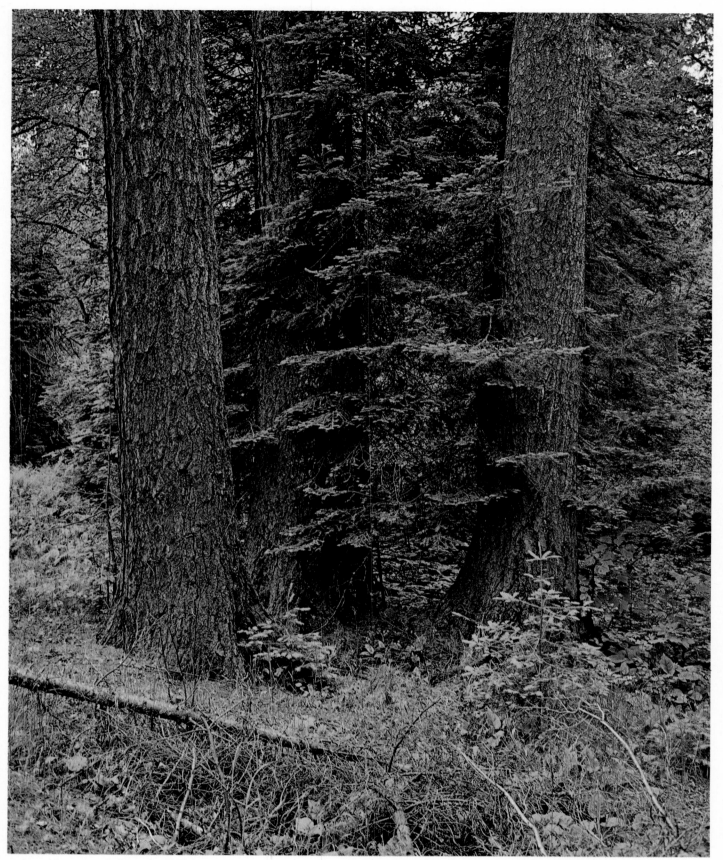

Vulnerable to fire, balsams crowd fire-resistant 100-year-old white pines; without a blaze, the firs may well survive the aging pines.

A Record of Ancient Fires

The north woods are full of sharply defined demonstrations of the pattern of fire's impact. In a grove of red pines on an island in Basswood Lake *(right)*, where forest colossi 100 or 200 years old tower above the canopy, the periodic passage of flames can actually be dated. These great trees bear the scars of past fires; by boring sections through the scars and then counting the growth rings around them, scientists can produce a calendar on which the dates of ancient fires can be exactly calculated.

In other places where fire has recently passed, a brilliant swatch of reddish-purple fireweed frequently blooms, prospering in the soil of wood ash. Slightly older burns will be dotted with the light green of birches and aspens or the gray green of jack pines, and those older still with the deeper hues of red or white pines. Areas unmarked by fire for many decades will be revealed on the terrain by the dark, coniferous growth of spruces and firs, which become dominant only when there are long periods between blazes.

Increasingly aware through such observations of fire's crucial role in nature, ecologists are re-evaluating the fire-prevention programs in the north woods and other wilderness areas. Although fire can have a devastating effect in the far north, it is controlled burns, some suggest, that might help the preservation of the rich forest along the southern limits of the north woods.

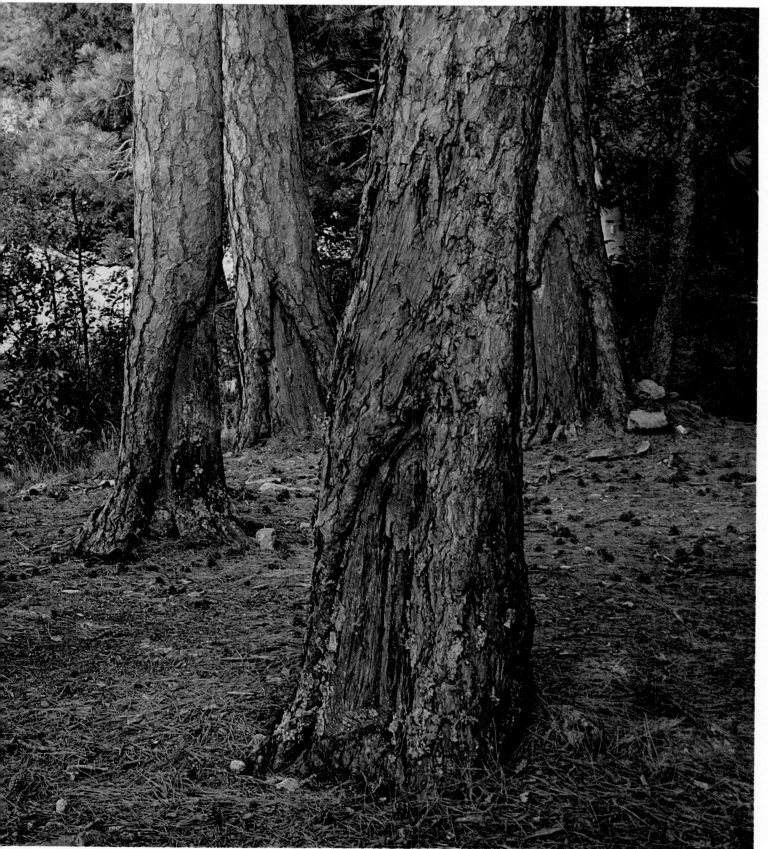

Fire scars almost 100 years old mar 200-year-old red pines. At far left is a relatively young, unmarked offspring of one of the hardy giants.

2/ The Formation of a Landscape

...this anomalous land, this sprawling waste of timber and rock and water...this empty tract of primordial silences and winds and erosions and shifting colours.

HUGH MacLENNAN/ *BAROMETER RISING*

The wind is cold. The countless pools in the potholed rocks shine dimly. Herring gulls, pintail ducks and Canada geese fill the air with mournful cries. I am 800 miles northwest of Grand Portage, not far from where the boreal forest dwindles out in wind-battered fingers of black spruce and tamarack, near the place where the broad and powerful Churchill River empties into Hudson Bay. On cloudy days from this point the sea is like iron beneath an iron sky. This is the southern edge of the Arctic tundra, a stark, bleak land where winter rules pitilessly for more than half of the year.

Here, where the great, shadowy forest reaches its northern limits, I sit on an outcrop of rock on an autumn day and brood on the history of this ancient land. How different it is here from the sparkling lake country of northern Minnesota and the Quetico-Superior region! That is a land that lives vivaciously in the present, gay and beautiful, alive with the song of the white-throated sparrow, the wild cry of the loon, the peaceful labor of the beaver; a land in whose limpid river waters one can see the dark shadow of the northern pike, the swift rush of the walleye, the darting of trout over sun-dappled rocks on the bottom. Here on Hudson Bay, everything is instead evocative of the past; the very crying of the birds overhead seems an echo from a time long ago that no man ever saw. This is a place where geologic time is a reality, where 10,000 years are but a breath in the earth's slow pulsation, where the long history of this lonely land can be seen, felt and understood.

Hudson Bay, on whose shore I sit, is the remnant of the last of the great inland seas that successively inundated the north woods. The seas that preceded it disappeared hundreds of millions of years ago. The worn ledge that holds me is composed of the compacted remains of myriads of aquatic organisms that built up layers of limestone in those ancient seas. Through countless seasons waves pounded my ledge; rain lashed it; snow and ice covered it; Arctic temperatures cracked and split it; summer sun beat down on it; and now here it is, a listening post on a tumultuous past.

But my ledge is not really very old, perhaps only 425 million years or so. Within a day's hike from here I can find granitic outcroppings of far more ancient stuff—some of the oldest rock in North America, formed 1.7 billion years ago, when life was still in a very primitive stage. And even this is not the oldest rock to be found in the north woods. Far to the south of Hudson Bay, on Burntside Lake near Ely, Minnesota, there is a point of land composed of rock that is more than 2.5 billion years old —greenstone, it is called, and appropriately so. It is dark green in color, and compact and solid as only a permanent buttress of the earth can be. It was originally formed of lava spewed forth from great fissures and fractures in the earth beneath the inland seas. Greenstone is found in many areas of the north woods—one two-mile-wide band of it in northern Minnesota stretches from Tower through Ely to Moose Lake, a distance of 40 miles. In places, this greenstone is 20,000 feet thick.

In close proximity to the Ely greenstone lie deposits of the rich iron ore of the Mesabi and Vermilion Ranges that brought untold wealth to those who discovered it in the last century and gave rise to such towns as Ely and Tower and Soudan and Virginia. Near the Pigeon River these deposits are so concentrated that they brought confusion to the compasses of early explorers and voyageurs, and still confuse the canoeist today; Magnetic Lake is named for this phenomenon. The name of Gunflint Lake in the same region bespeaks a different geologic inheritance. Flint, a sedimentary mineral that is yet another leftover from the inland seas, occurs here in such abundance that both Indians and white men named the lake to fix the spot where they could find the spark-producing stone that could be flaked into sharp tools.

All of these different types of rocks, and the myriad minerals of which they are composed, share a common heritage. All of them are part of the Canadian Shield, an ancient geologic formation that, in one area, has rock 3.8 billion years old—only a billion years or so after the earth itself came into being. Composed of the eroded roots of some of the earliest mountain chains that reared themselves above the earth's primal crust, the Shield reaches in a vast ellipse across the upper tier of the continent from the Atlantic to the Arctic Sea in the northwest, underlying 1,864,000 square miles of Canada and the Lake Superior and Adirondack regions of the United States. The Shield is thus the very lap of the earth to much of the north woods, and for the camper it offers particular advantages: wherever it crops out, excellent campsites are to be found, properly elevated for good drainage, often with fine ledges of the granitic rock of which the Shield is primarily composed. On them, one can draw up a canoe, build a fire and sit comfortably before a tent pitched on level ground.

Only about one tenth of the ancient rock of the Shield is exposed, but the Shield is very close indeed when one builds a campsite in the fragrant shade of spruce and pine and balsam. The soil of the north woods, the earth that nurtures all of its plants and trees, is seldom more than 10 inches deep; just below lies the Shield itself. That is a very thin covering for a place where trees grow 100 feet or more tall, and the effect of it can be seen on any tree, great or small, that has been toppled by a storm. No thick taproots probe deep into the earth and help to anchor the giant above; instead there is a tightly woven mat of roots and rootlets extending in all directions from the tree trunk. On ledges where the soil is especially thin, some of the rootlets find their way into cracks and fissures of the bedrock; often, as the tree falls, it pulls chunks of crumbling rock up with it. In such places the mat of the root system, peeling from the rock below, may lay bare a large area that has never reflected the sun. Once, on the Granite River north of Gunflint Lake, I slept in the shelter of a root mat: the rock beneath my sleeping bag was almost a pristine white, so cleanly had the earth been peeled back when the tree fell. I slept that night with my face turned upward toward the clear, bright stars, and it was as though I were a part of the earth, and had grown up here with it, a living limb of that ancient rock beneath me that now, once more, lay open to the world.

But man is a latecomer to the north woods. The earliest time at which he might have inhabited them is thought to have been some 14,000 years ago; more likely, the first man wandered in around 8,000 years ago. Whenever he did come, he remained a part of the forest ecology for thousands of years, living in harmony with it until the first white men arrived. In the main, the north woods Indians did not progress be-

The contours of a limestone ledge on Hudson Bay's western shore reflect millennia of polishing and erosion by glaciers, sea and rain.

yond the Stone and Copper Age until they were seduced by the wares brought from Europe by fur traders—and until they were corrupted by the white man's firewater, which undermined the Indian's mind and body, and his pride.

The north woods could not be inhabited at all, however, before the glaciers began to recede from the land 15,000 years ago. And it was the glaciers that made the north woods what they are today. These great ice sheets formed during the Pleistocene Epoch, which began perhaps two million years ago. The Canadian Shield was relatively stable by that time: its period of mountain building had long since ceased, and the inland seas had disappeared. The continent's outlines were generally similar to those of the present day, but the surface was in many respects very different. Most of the familiar geographical features, such as lakes and rivers, either did not exist at all or appeared in different forms. There were, for example, no Great Lakes to the southeast of the north woods; where these are now there were five great river systems. There were very few lakes in Minnesota and Canada, where they are found in such numbers today. Instead, there was a thick covering of forest, interrupted only by patches of bogs and grassy meadows.

This landscape was to be greatly altered by the onslaught of the Ice Age. For millions of years before then, the world had been growing colder. The temperature dropped only a few degrees in all, but in northern Canada that was enough to make the winters longer and more severe. Snow fell in masses over increasingly protracted periods of time; at last it fell nearly all year round, more snow than could possibly evaporate or melt and flow off in the increasingly short weeks of summer. And as it fell, it built up in the Canadian northland, layer by layer.

As time went by, the snow line advanced southward into the temperate zone. But far more ominous was the fact that the snow gradually turned to ice. As the snowfalls piled up, one on top of the other like the sediments in ancient seas, they pressed down and compacted the bottom layers—metamorphosing them, to use the geologists' term. When the snow was thousands of feet deep, the weight of it was enormous. The bottom layers finally were transposed into a blue-white ice that was, in effect, a rock with a melting temperature of 32° F. And then the great ice sheets started to move, adding to their own mass the accumulation of rock and gravel that they picked up and carried with them.

The movement of ice is a force difficult for the mind to comprehend. It is slow—advancing from an inch to 10 feet a day in most cases—but its power is inexorable, relentless and absolutely irresistible. Only two things in the world can stop it: warm weather or the ocean water off a continent's shores, where the ice perforce breaks off and sails away in the form of icebergs. Everything else succumbs; hills, valleys, lakes, soil cover, gravel, rocks, forest are all altered by the pressure and movement of thousands of square miles of ice thousands of feet deep.

The basic force that drives an ice sheet is gravity. As the snow falls year after year it accumulates more deeply in some places than in others. It also continuously turns to ice and ultimately, under the steady pull of gravity, the bulging edges begin to move outward in all directions. Ice moves when it reaches a depth of more than 300 feet, and because of its great weight—about a third as much as rock—it bends the earth beneath it as it goes. In this fashion Hudson Bay was formed during the Pleistocene Epoch by the great Laurentian ice sheet that covered all of northern Canada and ultimately advanced deep into the northern lands of the United States. Thus the ice shapes the earth not only by scraping, gouging and pushing material before it as it moves along, but by actually depressing the surface of the earth as well: its action might be described as that of a combination bulldozer, road scraper and steamroller, all in one unimaginably massive machine.

The glaciers of the Pleistocene Epoch—there were four major periods of glaciation, each lasting 100,000 years or so—scraped Canada almost bare. Forest, soil and rock debris were bulldozed away. The finer materials were picked up by the wind and deposited for the most part in what is today the American Middle West, where they formed the fertile fields of its farm states, extending as far south as Louisiana.

The marks left by a glacier's passing are everywhere to be found in the north woods today. Some of the evidence is small enough to be readily seen and understood. The Granite River, for example, has numerous islands and ledges that clearly show the planing and plucking action of the glaciers that passed this way: where the ice rode up on a rock ledge the surface is polished smooth, while at the other end, where the glacier froze to the ledge before moving onward, the rock is broken and split and in its steep, clifflike face gives evidence of how the ice tore off great chunks of it and carried them along. Much of the tumbled landscape of the Granite River area is due to glacial action. Everywhere can be seen debris deposited by the glaciers, from small pebbles to large, rounded boulders.

In some parts of the north woods, rock ledges show deep parallel groovings, like the claw marks of some ancient and gigantic bear. This

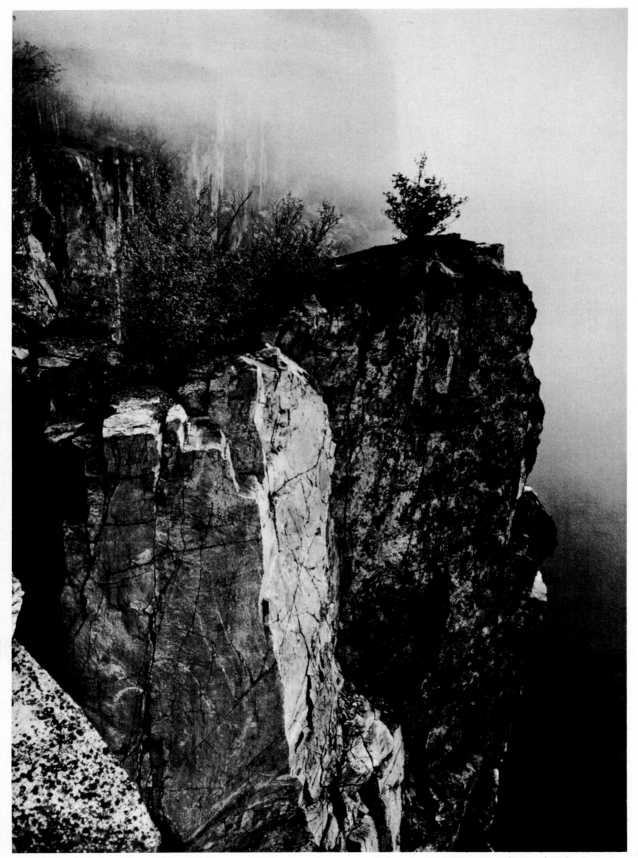

Precambrian lava, spewed from the earth a billion years ago, formed this sheer cliff at Palisade Head, Lake Superior.

is so, for example, on the dark greenstone point at Burntside Lake. The grooves were made by the scraping action of the rock-bearing glacier that moved across the greenstone ledge perhaps 15,000 years ago. And higher up on this same point—as on countless others throughout the north woods—are boulders called "erratics," which were left when the glacier melted and retreated northward. Some of these erratics are as large as houses and can be traced to their point of origin, which may be several hundred miles to the north.

These are traces that can be seen by the canoe camper today as he pitches his tent by some smooth ledge. But the full impact of what the ice age did to the north woods region can be best understood when one views the area from an airplane, or looks at a map that shows the thousands of lakes and rivers splashed across the land. For all of these glittering waters are the work of the glaciers. The channels of the rivers were gouged or hollowed out by the ice as it crept ponderously across the land. Some lake beds—kettle holes, these depressions are called—were created when enormous blocks of ice broke off from the main body of the glacier and were left stranded amid or beneath quantities of material also left behind when the glacier retreated northward; when the ice melted, bodies of water were held trapped within the debris. Other lakes were formed when debris left by the glaciers dammed up a valley or depression, leaving, for the time being at least, no outlet for the melting water when the glaciers retreated.

As the ice receded, a fantastic landscape slowly emerged. Bare of soil and plant cover, part of the bedrock of the Canadian Shield of the north woods country reappeared, gleaming dully in the wan sunlight of the Pleistocene spring. Over thousands of square miles this barren landscape was dotted with clear lakes of the melted glacial water. No plants or organisms of any kind obscured these cold, crystal waters, for nothing had grown here for thousands of years. The lake bottoms glistened with granite pebbles and white quartz sand, as pure as the water that covered it. Immense boulders lay scattered about as though thrown in fretful impatience by a giant child. Only the water moved, and it moved everywhere, making outlets for the lakes and channels for the rivers, curling through valleys, brawling down rapids.

Much of what is today northern Minnesota, eastern North Dakota, western Ontario, southern and central Manitoba and east-central Saskatchewan lay beneath an enormous lake of meltwater that, sometimes expanding and sometimes contracting, spread at its greatest extent over

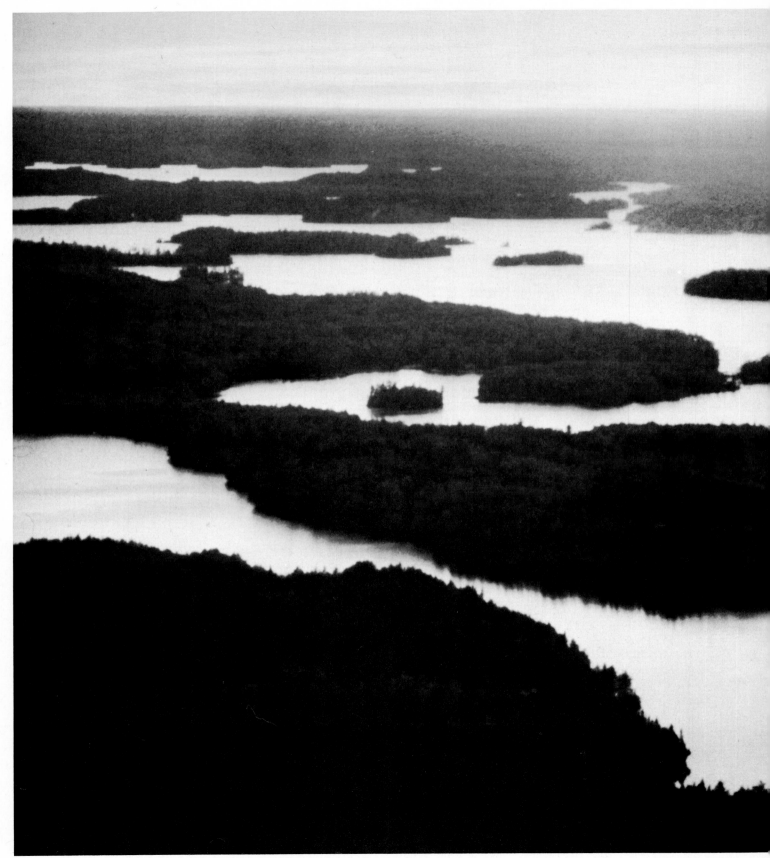

A prototypal north country panorama—dense woods of conifers interspersed with lake waters—unfolds in this view of Superior National

Forest. Such lake formations resulted when meltwater filled hollows dug by glaciers, or when drainage backed up behind their debris.

an area 700 miles long and 250 miles wide. Here the enormous weight of the icecap had depressed the earth's crust, leaving an immense watery basin. This lake, which geologists named Lake Agassiz in honor of Louis Agassiz, the 19th Century Swiss-American naturalist who first postulated the ice age, drained off slowly as the earth lifted again. What is left of this lake basin is now Lake Winnipeg and the lakes west of Quetico Provincial Park in Ontario.

Reminders of long-gone landscapes are all around me as I sit on my ledge of ancient limestone on the western shore of Hudson Bay. The cold wind blows down from the ice sheets of the Arctic as it did almost a million years ago. There are no trees; the forest has not returned here. This is a primeval world; if I did not know that I was looking out at the gray expanse of Hudson Bay, I might well imagine that before me lay the waters of Lake Agassiz.

I am, in fact, in the middle of a perfect model of what the glaciers left behind—and also of what followed after their departure. The very rock on which I sit was shaped and polished by the glaciers. The ice scratched the long, parallel grooves in it. Turbulent water scoured its potholes; in them, in turn, are particles of sand and rock left by the glaciers. On such a barren landscape the Pleistocene sun shone 10,000 years ago when the last glaciers left the north woods. What a sight that must have been! Hundreds of square miles of scraped and scoured rock, much of it covered by glacial debris, stretching farther than the eye could see—a reborn world. And here before me, on the shore of the great cold bay, lay its duplicate.

To the south of Hudson Bay, life returned to this reborn world with the spores of lichens floating through the air, borne by the ceaseless wind. Where the spores landed on the sun-warmed rocks they found minuscule footholds and lichens began to grow. Red, rust-brown, gray, dark green, flaming orange, burning yellow, they brought color, and life, back to the dead land.

Lichens are marvelous plants. They look like splashes of paint on the rocks, clinging as though they had dried there. But they are strong, unbelievably strong, and hardy. A combination of two primitive plants, fungi and algae, they were indispensable for the re-creation of these glaciated regions, for they can grow where nothing has grown for millennia, and they contribute to the production of the substance that advanced and complex plants need to grow on: humus, the decomposed residue of plant tissue, the life stuff of soil. The fungi and algae

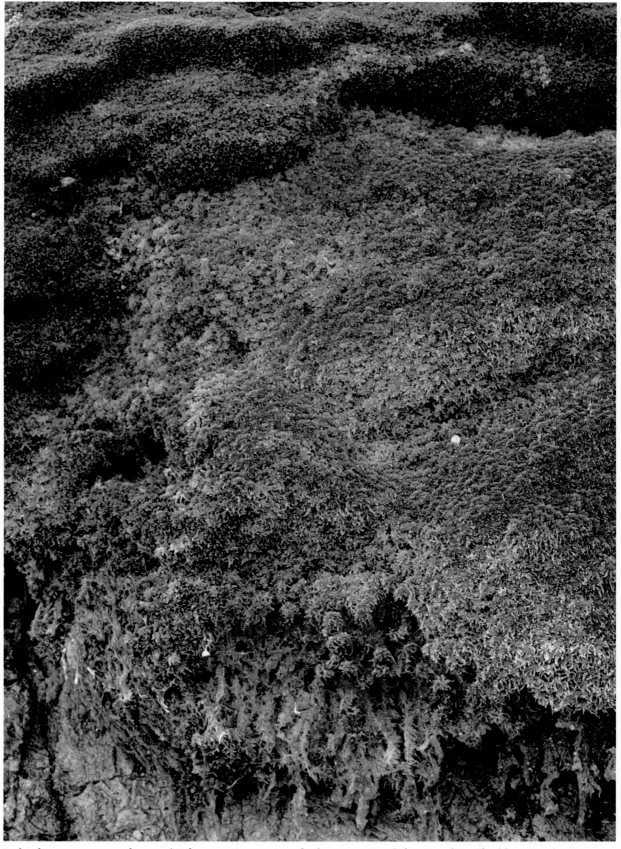

Red sphagnum moss and green feather moss grow atop the bare granite of the Canadian Shield near Lake Superior.

of lichen live symbiotically, each to the benefit of the other. The fungi draw some minute sustenance from the rocks and provide moisture to the algae, while the algae manufacture sugars—upon which the fungi feed—from the energy of the sun. And as the lichens spread, they help to break the weathering rock down into finer particles that seep into cracks and fissures and eventually play host to other plants.

Some of the weathered particles also fall into the pools of water in the rocks, adding to the material at the bottom. Here, in time, water plants take hold. Meanwhile, on the rocks the moss grows, adding its own remains to the slow accumulation of humus until grasses can germinate; then come plants like the fireweed, and finally the tiny beginnings of a tree.

Gradually, after the glaciers were gone, the soil accumulated atop the rocks of the north woods. More mosses grew and then seeds drifted in, carried by birds or by the wind. The grasses came, and the herbs and shrubs, and the familiar birches, willows and aspens. Berry bushes arrived and prospered in a profusion that beckoned to animal life. Then came the lowly, hardy jack pines, the black spruces and the white, and finally, in the southern reaches of the woods, the towering white and red pines that would eventually dominate the forest until the axes of the white man felled them in their prime.

And gradually, too, there came the animals, migrating northward as the plant cover grew and proffered the foods they needed. They were a different lot from the beasts that had been there before the ice came. They were also different from the animals of the ice age itself, like the mastodon and mammoth; most of these had almost gone, and many of their carcasses were entombed in ice or frozen swampland. Gone, too, were the camel and the tapir and other ungulates, and the great, saber-toothed cats that had preyed upon them. Some of these animals had traveled southward ahead of the ice and eventually reached the southern continent, where some of their descendants survive. But the beaver appeared in the northland, and the white-tailed deer, to feed on the tender shoots of the young deciduous trees; and the moose came to feed on the water lilies and other aquatic plants of the lakes; and the wolf came to prey on the moose and the deer, and in time there came the wolverine, and the mountain lion and the black bear. And so in time the north woods became populated with the animals that are there today.

The environment of the north woods—the forests, the lakes, the bogs —is very new in geologic terms, despite the great age of the rocks on which it is based. The 10 inches of soil that today cover the bedrock took 10 millennia to accumulate; as geologic time is counted, this is brief, a proof of how quickly nature can recover after a disaster.

And yet this land, like all lands, carries within itself the seeds of its own destruction. The disappearance of the north woods lakes, for example, is as inevitable as the seasonal ebb and flow of the water in their basins and in the streams that feed them. If they are placid lakes with little current, they gradually fill up with sediment; plants take over, moving outward from the shore. Eventually the lake becomes a bog, and the bog in time fills up with soil and the forest takes it over. There are innumerable places in the north woods where one can see every stage in this process—from a bog in the making to a bog taken over by the trees.

If the lake has a current, the water at the outflow will gradually erode the threshold, deepening it until at last the whole lake will drain out, leaving behind a river. If the climate turns dry, the lake will evaporate faster than it can be refilled. Born of geological accidents, lakes may die from them too: an earthquake, nearby or distant, may change the drainage pattern of an area so that the lake will lose the sources of its replenishment.

Such a process, of course, may take thousands or even hundreds of thousands of years. Or it may take fewer. The floor beneath Hudson Bay, originally depressed by the glaciers, is gradually uplifting now that the enormous burden of ice has been removed. The shoreline is reaching into the water at the rate of two feet a century, and geologists calculate that within 40,000 years the entire 478,000 square miles of the bay—an area in which Texas and a few other good-sized states could be comfortably accommodated—will be reduced to a tiny inlet. In time, the forest will encroach upon parts of the former sea bed. But by then there may be still another ice age to bury the land, the lakes, the woods and the works of man and start the cycle anew. Thus the north woods, in all of their various aspects, are an affirmation of life—for through the long ages of their geologic history they have died many times, but have always come back again.

3/ The Voyageurs—a Legend

Paddling and portaging their way westward...where none but the hardiest could survive, the highhearted voyageurs...opened Canada's rich hinterland.

HUGH MacLENNAN/ *BY CANOE TO EMPIRE*

In the history of the north woods there are two undisputed heroes: the legendary logger Paul Bunyan and the French-Canadian voyageur. Their likenesses greet visitors in many roadside places: Bunyan, shown 10 or 15 feet tall, straddling a huge pine log atop some filling station or motel, arms akimbo in his checked mackinaw, face split in a gap-toothed grin; the voyageur, small, stocky, hunched under a backbreaking load as he is pictured trotting down some portage trail or paddling along some wilderness waterway, his mouth open in full-throated song. They are the very incarnations of the woods, these two, yet they stand for entirely different things.

Bunyan's legacy is a wilderness that is drastically changed from the one he knew. The lumberman's ax, which his storied army of loggers wielded with such efficiency, rang through the Minnesota woods from the Pigeon River to Rainy Lake, and from there on into Canada; and where Paul's giant foot trod, there the land was devastated. Large-scale logging began around 1870 and continued unchecked for about 50 years, until it tapered off for lack of prized red and white pine as well as by reason of a belatedly awakened public conscience. John Szarkowski, in his book *The Face of Minnesota,* sums up what Bunyan symbolized:

"In the mid-nineteenth century Minnesota contained thirty million acres of virgin timber. The cutters began slowly; four thousand were in the woods in 1870, forty thousand in 1900. The peak year was 1905:

nearly two *billion* feet were cut, 98 per cent of it white pine. Ten years later the yield was half as big; twenty years later, a fifth. Then the forests were gone. Remaining were the tales of Paul Bunyan, who leveled a section of pine when he sneezed."

The legacy of the voyageurs is quite different from that of the loggers. The voyageurs were content to leave the woods as they found them: dark, austere, abounding in mystery. Despite the hardships they endured, this was an environment they loved. Their attitude showed in their relations with the Indians, with whom they generally lived in friendship. "They accepted the Indian at his own evaluation, which was not low," writes the historian Grace Lee Nute, "whereas the American frontiersman could scarcely find words for his contempt of what he considered a thieving, shiftless, dirty race." Many of the voyageurs intermarried with the Indians and raised families of half-breed children that formed the backbone of the early north woods settlements.

The voyageurs were a class of men as distinct in their time as loggers were in theirs. They have sometimes been confused with those other romantic figures of the north woods, the *coureurs de bois,* or woods runners; but these were freelance trappers and traders who worked their traps alone and traded in furs without a government license. The voyageurs were *engagés,* hired canoemen who signed on for a one- to three-year tour of duty, after which, more likely than not, they would sign right up for another. They were paddlers, not traders; they had no interest in the fur trade nor any ambitions to be rich or successful in anything but meeting the endless challenge of the wilderness. They were as tough as nails, they gloried in risking their lives in rapids, and they had their own peculiar pride in excelling at a way of life that strikes us today as not much different from that of a Roman galley slave. Grace Lee Nute quotes one voyageur:

"For 24 [years] I was a light canoe-man. No portage was too long for me. I could carry, paddle, walk and sing with any man I ever saw. When others stopped to carry at a bad step I pushed on—over rapids, over cascades, over chutes; all were the same to me. . . . Were I young again, I should glory in commencing the same career. There is no life so happy as a voyageur's life, none so independent; no place where a man enjoys so much variety and freedom as in the Indian country. *Huzza! Huzza! pour le pays sauvage!*" And this from a patriarch who still, in his seventies, longed to paddle.

The life of which this old man spoke so fondly was almost inconceivably rugged. Traveling in birch-bark canoes that had to be handled as gently as eggs, squatting or kneeling among the bales of freight, the voyageurs paddled for 15 to 18 hours every day, stopping only for five minutes or so at the end of every hour to have a quick puff at their pipes. They usually ate but two meals—breakfast, after three or four hours of predawn paddling on an empty stomach, and supper, often as late as 10 o'clock at night, consumed by the light of a campfire. The voyageurs had scant protection against the swarms of black flies and mosquitoes that tormented them day and night, nor were they shielded against the weather. They paddled in pouring rain, high wind, burning sun; only snow and ice could limit their exertions.

But the most trying of all the hardships the voyageur endured were the portages, the overland trails they had to traverse between navigable stretches of water. The ordeal began with the unpacking of the canoes. Because of their fragile birch-bark skins these could not be dragged up on the bank for unloading, so the men had to leap out into water that was often waist deep and carry the bales of cargo ashore. They would then sling the bales onto their backs and begin the arduous journey overland. And arduous is the word for it: each voyageur was expected, as a matter of routine, to carry at least two of the standard 90-pound bales on each portage trip, and often he carried three. Daily the men ran the risk of hernia from their burdens; more than a few of them died of this injury on the trail.

To help in carrying their loads, the voyageurs used a tumpline, a device adopted from the Indians. This was a broad leather strap, about three inches wide, that passed around the forehead and over both shoulders—bringing the powerful muscles of the neck into play—and down the back. The bottom, or anchoring bale, would be tied with smaller straps attached to the tumpline. Slung into the tumpline on top of this bale, along the carrier's spine, would be placed as many more bales as he could take—perhaps two more, for a total of 270 pounds. Stooped under this towering mass, the voyageur was always, it seemed, on the verge of pitching over forward. Nevertheless, he would trot rather than walk, and at a phenomenal pace. One passenger on a canoe journey, a missionary on his way to an outpost in the woods, described how he had tried to keep up with the crew on a portage. "I *ran* faster than I chose," he recalled. When a voyageur reached the end of the portage, he would set down his load and immediately trot back to the start of the trail for another. If the portage was more than half a mile long the cargo would be moved along it in stages.

A late-19th Century voyageur, in working garb typical of the period, was sketched for the magazine Harper's Weekly by the American artist Frederic Remington on a north woods trek.

Meanwhile, the canoe also had to be carried over the trail. The Montreal canoe, used between Montreal and Grand Portage, weighed 600 pounds; the North canoe, used between Grand Portage and the north, weighed 300 pounds—in either case a formidable proposition. Aside from its weight, the canoe was clumsy to carry, yet required the most delicate handling. The bowsman and the steersman of the North canoe took charge of this task, carrying the craft upright. In the case of the Montreal canoe, it was borne bottom up by four men, their heads inside it, trying at the same time to see where they were going. Not long ago, portaging along the Pigeon River, a friend and I experienced firsthand the problems of carrying a canoe through thick woods. We had come ashore at a place where a well-marked trail appeared to lead inland from the muddy banks of the Pigeon. But we sank ankle-deep into mud at every step and within a few yards we found ourselves in a trackless tangle of fallen trees and willow and birch shrubs. While I scouted ahead to find the direction of the trail, my friend, carrying our aluminum canoe on his shoulders as if he were some giant silver turtle, stumbled along behind, unable to see anything except the ground immediately before his feet. As a result, he progressed in a series of loud bonging noises that rang through the woods as our canoe rebounded from tree after tree. It was a noise fit to wake the dead, though such was not our intention. A birch-bark canoe would have been smashed to ribbons; our craft, made of metal, survived. Later we decided that we had been misled by a beaver trail; the portage we were looking for lay on the other side of the river.

The voyageurs, of course, often had to scout and hack a trail out of the wilderness by themselves, and even on known trails the perils were numerous. Some portages were fairly easy, leading over level or gently rising ground. But other portages might lead through bogs or seemingly bottomless mud or deep sand that tugged at the feet with every step. Always there were rocks and roots on which one might stumble and perhaps fall with an ankle-twisting wrench. Some trails led over cliffs, where the only foothold was a series of steplike ledges. In all, along the 3,000-mile route of the fur trade from Montreal to Fort Chipewyan on Lake Athabasca, there were 120 portages to be negotiated, and every one of them required a sharp eye and a sure step.

The most notorious was the 12-mile-long Methye Portage, in what is now the Canadian province of Saskatchewan, between Methye Lake and the Clearwater River, not far from the voyageurs' northern terminus at Lake Athabasca. Peter Pond, a Yankee fur trader whose nasal

Connecticut twang still echoes through his phonetically written journals, was the first white man to discover this portage, and many were the voyageurs who wished he never had.

Most of the portage is innocent enough. Its first eight miles—still well marked today—lead in a fairly direct line across a comfortably smooth, sandy ridge, sparsely wooded and with no sharp elevations to surmount. Then the portage is interrupted by a small, sparkling gem of a lake—Rendezvous Lake, only a mile long. Here the voyageurs had to reload again, paddle across and unload for the second stage of the portage. Even so, Rendezvous Lake provided a breather; white sandy beaches border it and make excellent camping grounds. The second stage of the Methye Portage, however, offered the voyageurs no comfort. It is deceptively easy at first, climbing gently through rolling hills toward the Clearwater River, four miles away. But then, like a cornered beast, it suddenly turns vicious. Near its end, among hogbacked ridges and steep eroded gullies, the trail seems almost to disappear: it takes a precipitous 700-foot drop down a cliff.

From the top of the cliff the eye can see 30 or 40 miles down the narrow valley of the Clearwater River. The view is a memorable one. The great Alexander Mackenzie, the fur trader and explorer who was the first to complete an overland journey across the entire northern continent, reported that the cliff commanded "a most extensive, romantic, and ravishing prospect." But among the voyageurs, there were few who really appreciated it; their Herculean task was to get canoe and cargo down this cliff. Sleds were devised to cradle the canoes so they could be let down the precipice without damage. The cargo was let down on ropes, or slid along with the voyageurs as they scrambled down. Staggering as the job was, its difficulties were infinitely compounded on the voyageurs' journey southward from Athabasca to Grand Portage, when everything had to be hauled up the cliff.

Understandably, the voyageurs preferred to avoid portaging wherever possible, and in order to do so they often risked a more deadly wilderness peril—running rapids. The gamble was a tempting one. Successfully pitting a canoe against even a brief stretch of wild water could save hours of sweat involved in a transit overland; besides, it held a certain zest that was wholly absent from the drudgery of portaging. There were some rapids that headquarters in Montreal flatly forbade the voyageurs to attempt, for the men in the home office included former *bourgeois,* traders who had accompanied the canoe brigades and who well knew the dangers that rushing waters posed for human life—and for precious cargo. But the voyageurs often failed to heed these instructions. They took the risk and the consequences, shouting happily if they completed the run, burying their dead somberly if they failed. Each gravesite was marked with a cross for the man lost, and there are riverbanks in the north woods where, during the fur trade era, as many as 30 crosses stood clustered near the fast water.

If at all possible, the rapids were shot with every man and all the cargo intact in the canoe—the most timesaving though also the most dangerous method. Every man's life depended on the close and un-failing coordination of the bowsman and the steersman, and on their skill at skimming along between the eddies near the shore and the tur-bulent waves at the center of the rapids. Sometimes the *milieux,* or mid-dle men, paddled too, for it was imperative to keep the canoe on course.

The task was considerably more complicated in the case of upstream rapids. To negotiate them, several methods could be tried. One was to "track" the cargo-laden canoe along the rapids. All of the men, except the steersman, would scramble to shore and tow the canoe along by means of a long line to which they harnessed themselves like canal hors-es; where the shore was difficult to walk along they had to wade in the icy water, struggling to keep their footing among the slippery rocks and rounded stones of the river bottom. Another method, employed where the stream was fairly shallow, was poling; the voyageurs stood in the canoe and pushed it with 8- or 10-foot poles. In particularly rough spots, the men might unload half the cargo on the riverbank: the half-empty canoe would then be paddled or pulled up the rapids, unloaded at the top, then guided down through the fast water to be loaded with the rest of the cargo and moved up again.

Sometimes in shooting rapids the canoemen escaped death against all odds. Alexander Mackenzie gave an account of one such incident that befell him and his crew of voyageurs. They had encountered a fear-some stretch of water that forced them to hack a portage trail out of the woods to circumvent it. This done, they embarked but found them-selves almost immediately facing another bad rapid. In order to lighten the canoe Mackenzie got out, but his men persuaded him to get back in. Within minutes, however, they were in deep trouble.

"We had proceeded but a very short way," Mackenzie's diary re-counts, "when the canoe struck, and notwithstanding all our exertions, the violence of the current was so great as to drive her sideways down the river." Mackenzie and the entire crew leaped into the water and at-

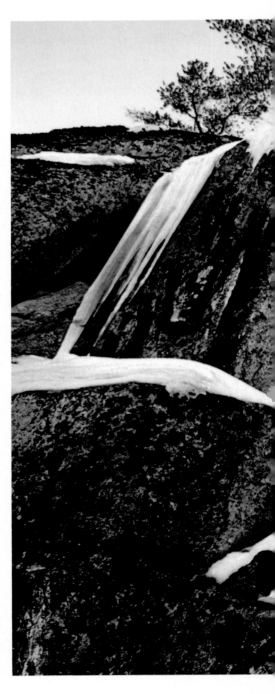

Pictographs painted by early north woods Indians, on a cliff face at Hegman Lake, Minnesota, include a godlike figure and forest animals. The artists used the earth's own pigments to adorn the sheer granite walls.

tempted to halt the canoe's downstream rush, but they could not keep their footing and clambered back in again. Meanwhile the stern had been shattered against a rock, and seconds later the bow met the same fate when it was driven into the rocks of the opposite shore. The bowsman grabbed at a tree to try halting the canoe; the tree bent almost in two and before he could loose his grip, he was catapulted in a great arc from the canoe across the foaming water to shore.

While the rest of the crew sat paralyzed and helpless, the canoe careened down the river, with sharp rocks ripping several huge holes in its bottom. Finally, when the gunwales were awash, the men went over the side again and managed to kick their way to an eddy where at last the downstream rush was halted. Here they were rejoined by the bowsman, who had made his way down through the woods. Surprisingly, the men were able to repair the canoe, but some days later, after several further spills, they had to abandon it and build another, which they did on the spot—in four days.

With very rare exceptions, the voyageurs were all small men; there was no room in a freight canoe for six-footers. Their faces were typically French: swarthy, thin, often deeply furrowed, quick to reflect emotion. Many of them also bore the stamp of rugged experience. A contemporary description of one crew tells of a man whose face "seemed to have been squeezed in a vise, or to have passed through a flattening machine; it was like a cheese-cutter—all edge." Another man had had one nostril bitten off in a fight: "He had the extraordinary faculty of untying the strings of his face, as it were, at pleasure, when his features fell into . . . a crazed chaos almost frightful." A third man had been slapped by a grizzly bear—"his features wrenched to the right."

The working dress of the voyageurs was highly distinctive yet utilitarian. They wore deerskin moccasins and deerskin leggings that reached to just above the knees and were held up by thongs attached to a breechcloth. The thighs and upper torso were left bare, although a short shirt of wool or deerskin was sometimes worn to fend off flies and mosquitoes. For special occasions, this basic uniform was embellished by such items as a red woolen cap with a tassel hanging down over one ear, a blue hooded jacket and a gaily colored sash from which hung the voyageur's short clay pipe and his tobacco pouch. Often, too, he sported a bright scarf that, tied loosely around the neck, also afforded at least the pretense of protection against stinging insects. Always before arriving at his destination, the voyageur would fancy

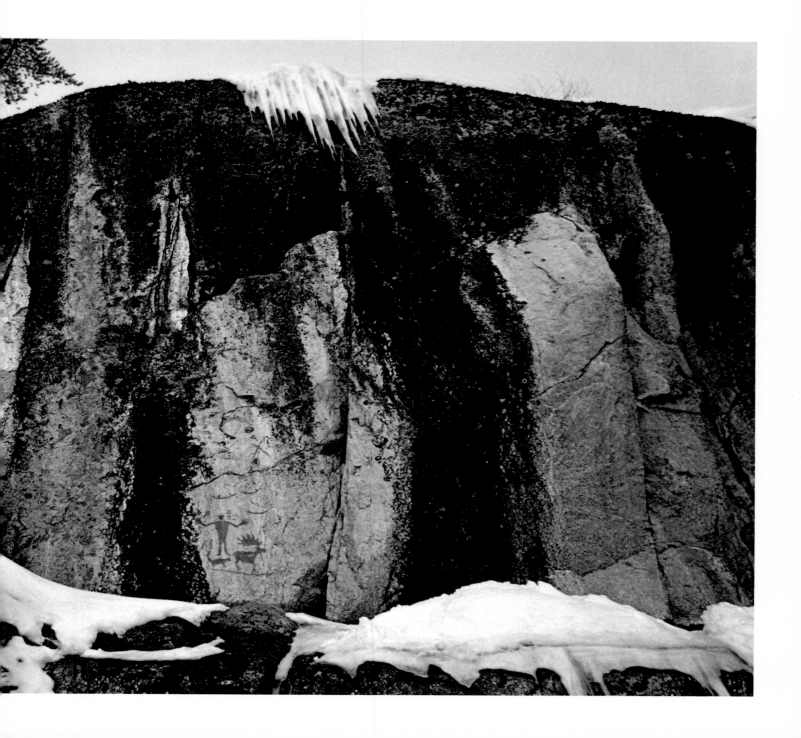

himself up with a few brightly painted or dyed feathers stuck into his cap—for the ending of a journey was reason to rejoice, a moment of triumph after long weeks of travail.

But en route there was little time or cause to celebrate. The voyageurs' working day began well before dawn, usually around three in the morning. "Lève, lève, lève!" would come the shout—"get up, get up, get up!" Groaning and cursing, the men would stumble about in the darkness until they had loaded and launched the canoes. Then they would set off, without so much as a morsel of food, and paddle for about three hours until daylight came and a stopping place was reached. The canoes were moored offshore by laying long poles across the gunwales with one end on land, and the voyageurs would fall to their breakfast, prepared the night before by the cook of the canoe brigade.

The voyageurs' menu was short on variety but long on quantity and nourishment. For the men who traveled the Montreal-Grand Portage route, the basic dish was a mushy soup composed chiefly of dried peas or corn, boiled up in a big iron cauldron with water from a stream or lake until the mixture was thick enough for a spoon to stand upright in it; and since these men were the so-called pork eaters, the soup might be flavored with a little salt pork. For the winterers—the men of the north—the main food was the Indians' staple, pemmican, a mixture of dried buffalo or moose meat and berries, pounded fine; sometimes the pemmican was cooked in a soup of water and flour, a concoction called rubbaboo. And if the voyageurs caught any fish or chanced to shoot some game in their brief periods of respite from the paddle, these too would be joyfully added to the pot. Occasionally there would be a disappointment in this regard. One journal records that on a summer day in 1775, on the Pinawa River above Lake Winnipeg, "The mosquitoes were here in such clouds as to prevent us from taking aim at the ducks, of which we might else have shot many."

Sometimes the voyageurs' diet was augmented by the unleavened bread the French still call *galette*. In its simplest form, this bread was made by pouring a little water into a sack of flour; this was then kneaded until it formed a dough. If birds' eggs were available, they were poured in too. The dough was finally shaped into small, flat cakes that were baked in front of the campfire, or fried in a pan with grease. Campers today eat much the same sort of thing when they whip up a mess of bannock, the bread that is standard fare all over the north woods.

Whatever the voyageurs ate they ate with gusto, and in the most primitive manner imaginable. While some had dishes or cups, those who did not poured their ration of soup into their caps or kerchiefs, or into a shallow depression in a rock—in which case they would go down on their knees and lap up the food, dog fashion.

Once breakfast had been eaten, the men climbed back into their canoes and the heaviest work of the day began. Hour after hour they bent to their paddles, stopping only for the mandatory brief pause for a smoke, with no time out for a midday meal. On and on they drove, far into the evening. Thomas McKenney, an American who spent some years on the fur trade routes before he became an official of the United States Bureau of Indian Affairs, wrote of one voyage in 1826. At 7 o'clock one night, he asked his men if it were not time for the evening meal. "They answered they were fresh yet," he writes. "They had been almost constantly paddling since three o'clock this morning . . . 57,600 strokes of the paddle, and 'fresh yet!' No human beings, except the Canadian French, could stand this." When McKenney's men at last laid down their paddles, it was half past 9 and they had traveled, in the course of this one day, 79 miles.

The evening meal for the voyageurs was soup again—but not until the canoes had been unloaded and laid on their sides on the beach so that they could be inspected for leaks or weak spots. Each canoe carried a repair kit consisting of a ball of pine pitch, a roll of *wattape*—the tough, stringy rootlets of the spruce tree—and a sizable roll of birch bark, as well as some cedar planking in case the frame was broken. If a leak was found, it was repaired on the spot: pitch would be melted at the campfire, birch bark cut to cover the damaged spot; then the bark would be sewed on with *wattape* and the joint covered thickly with pitch. The men were experts at this work, which they had learned from the Indians; they repaired the canoes with a skill and patience born of the knowledge that their very lives depended on the care they used.

At night the men slept like the dead. Curled up under their canoes, they had no mattresses to ease the hardness of a rocky floor, and they usually had only one thin blanket each. Nonetheless, they slept so deeply that only a major calamity could have awakened them—all, that is, save the cook, who had to get up periodically to stir the soup he was preparing for the next day.

There was one aspect of the daily life of the voyageur that merits a special look: their habit of singing as they paddled along. Even as they set forth in the predawn dark they would launch into a rousing chorus that echoed incongruously through the pitch-black woods, and they would

continue to sing, hour after hour, throughout the day and into the evening. Their songs—gay and sad, moral and bawdy—were like magnets
pulling them on. Probably this custom of the voyageurs helped to relieve the tension of having to be ever watchful in the wilderness, and
certainly it had a practical use as well, for the rhythms of the songs set
a pace for the sweep of the paddles.

In any case, the songs were absolutely indispensable to the voyageurs, and the man with a good voice who could lead the chorus was
sometimes awarded extra pay. Some of the songs had been brought
over from France by an earlier generation; some were of the voyageurs' own devising. Composed in a French that is now archaic, the
songs are highly evocative of the singers' wandering lives. Many tell of
home and sweetheart and of lost love. One that is still familiar on both
sides of the border is "Alouette." Another, and certainly one of the love-
liest, is "A la Claire Fontaine" ("At the Clear Running Fountain"), which
became a sort of unofficial anthem of French Canada:

> *A la claire fontaine*
> *M'en allant promener*
> *J'ai trouvé l'eau si belle*
> *Que je m'y suis baigné.*
> *Lui y a longtemps que je t'aime,*
> *Jamais je ne t'oublierai.*

The song tells of a voyageur who stops by a clear running fountain
and finds it so beautiful that he bathes in it. As he dries himself beneath an oak tree, he hears a nightingale singing high in its branches,
with a heart as gay as his own is sad—for he has lost his lady love.
Why? Because one day he refused her a bouquet of roses that she de-
sired. Now he regrets his refusal bitterly. He knows that he will always
love her, never forget her, and he wishes that the roses that brought
about his downfall were thrown into the sea.

Other songs of the voyageur deal with the immediacies of the life he
has chosen, celebrating the virtues of his canoe, affirming his faith in
the power of his paddle, reminding him of the dangers that always
lurk. These are graphically summed up in "Quand un Chrétien Se Détermine à Voyager" ("When a Christian Decides to Voyage"), a sermon
in verse, supposedly given by a priest as the men prepared to embark.
The rhyming is lost in the English translation, but the import remains:

"When a Christian decides to voyage he must think of the dangers
that will beset him. A thousand times Death will approach him, a thou-

sand times he will curse his lot during the trip. . . . When you are on traverses, poor soul, the wind will come up suddenly, seizing your oar and breaking it and putting you in grave danger. . . . In the evening if the swarms of mosquitoes assail you unbearably as you lie in your narrow bed, think how this couch is the likeness of the grave where your body will be placed. . . . When you are in those very dangerous rapids, pray to the Virgin. . . . Then take the waves boldly and guide your canoe with skill. When you are on portages, poor soul, sweat will drip from your brow, poor *engagé*. Then do not swear in your wrath, rather think of Jesus bearing His Cross."

Such thoughts must have often been with the voyageurs on their long paddle northward. Not long ago I stood on the point of rock where Fort Chipewyan, the voyageurs' northernmost trading base, was established in 1789 on the shores of Lake Athabasca. I looked across that immense desert of water and marveled at the men who penetrated to this point in days when the closest contact with civilization was nearly 2,000 miles away.

Here in the far north, everything seems more lonely, more remote. The dark spruce spiral upward toward the sky, and the aspen leaves twinkle in the chill wind. Wide rivers run their swift course, and the land through which they pass is home to lynx and black bear, wolf and moose—some of the wildest country left on the North American continent. Looking out across Lake Athabasca, I felt the wildness and the loneliness. The sense of something not only untamed but untamable —this is what the voyageurs left for later generations.

A Proud, Hard Life

Churning rapids, trackless forests, precipitous cliffs, mosquito-ridden marshes—none of these obstacles could daunt the French-Canadian voyageurs who paddled and portaged their canoes through thousands of miles of north woods wilderness during the 18th and 19th Centuries. Outward bound, they carried goods to be traded to the Indians; on return trips they carried the prize pelts received in exchange—furs that enriched not the voyageurs but their employers: the North West Company, later absorbed by the Hudson's Bay Company.

In the course of their service for this English colossus the voyageurs came under the scrutiny of a remarkable woman who recorded their rugged way of life in oils and watercolors. Frances Anne Hopkins had little formal artistic training, but was the granddaughter of the renowned English portrait painter Sir William Beechey. At 21 she married Edward Hopkins, a Hudson's Bay Company official, and settled near Montreal in 1858.

Over the next decade the intrepid Mrs. Hopkins accompanied her husband on a number of wilderness expeditions, sharing the perils of the voyageurs' canoe brigades and their portages through the woods. From the sketches she made en route came a score of finished works that memorialized the rough-hewn faces and brawny figures of the voyageurs, the decorations of their birch-bark canoes, the ripples and reflections of northern waters. Among her works are the oils on canvas shown at right and on the following pages.

Mrs. Hopkins' pictorial narrative, more evocative than the written accounts of the life of the voyageurs, also served as an epitaph for their valorous era. By 1870, when she and her husband retired home to England, diminishing numbers of these stouthearted men were plying the swift, turbulent waters of the wilderness and tramping through its dark woods. After more than two centuries the fur trade itself had begun to enter a protracted decline, giving way to more lucrative ventures such as lumbering and mining. Improved, if less romantic, means of transportation came to supplant the birch-bark canoe: steam-powered boats and, in time, the railroad, whose ribbons of steel were laid down where once only footpaths existed. By the start of the 20th Century, the voyageur breed had totally vanished from the north country.

As the bowsman watches for dangerous rocks, voyageurs run a canot de maître, the so-called Montreal canoe, down a savage stretch of white water. The few passengers shown probably include the painter and her husband; the others are presumably Hudson's Bay Company officials.

A brigade of Montreal canoes moves off into the mist. In this work, which she entitled "Canoes in a Fog, Lake Superior," Mrs. Hopkins

> The voyageurs and travellers take their seats, a hasty look
> is thrown around to see that no stray frying pan or
> hatchet is left behind, and the start is made. An effort to be
> cheerful and sprightly is soon damped by the
> mist in which we plunge, and no sound but the measured
> stroke of the paddle greets the ear.
>
> HENRY Y. HIND/ NARRATIVE OF THE CANADIAN RED RIVER EXPLORING EXPEDITION/ 1860

and her spouse are seated in the nearest canoe. She has her sketchbook prepared, while Mr. Hopkins puffs at his pipe reflectively.

The pale light of evening reveals the lakeside camp of a company of voyageurs. Some gather firewood, others oversee the preparation of

Here we found waiting for the morn seven loaded
canoes and eighty voyageurs belonging to the Hudson's Bay
Company.... It was an uncouth scene. There was a semi-circle
of canoes turned over on the grass to sleep under, with blazing
fires near them, surrounded by sinister-looking long-haired
men, in blanket coats, and ostrich feathers in their hats, smoking
and cooking, and feeding the fires.   JOHN J. BIGSBY/ THE SHOE AND CANOE/ 1850

a meal, another scans the horizon, and one man, apparently exhausted after a day's paddling, slumbers beneath a canoe.

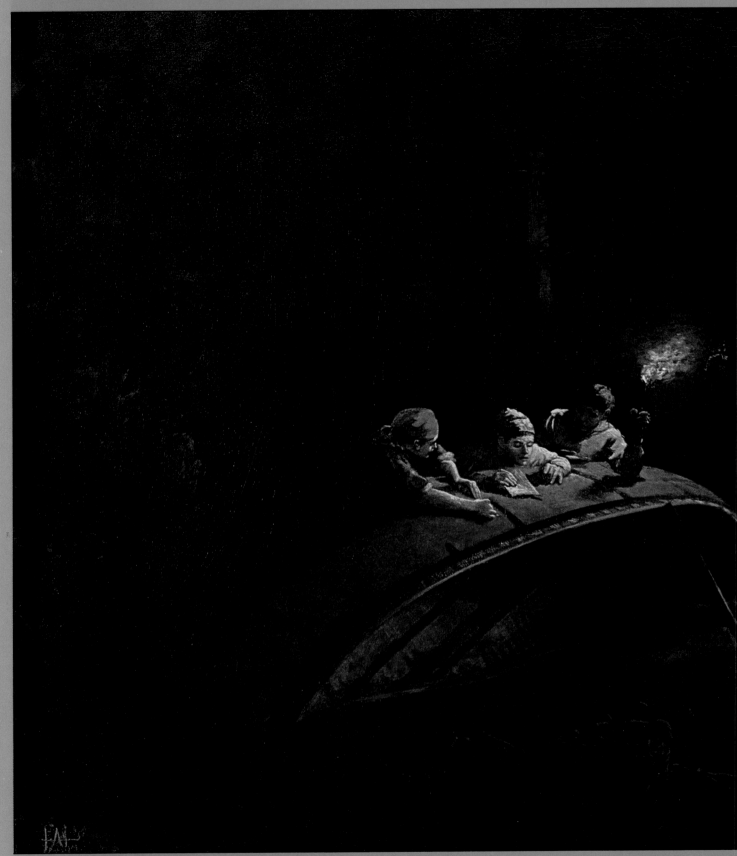

At night the vital task of repairing their canoe absorbs a voyageur crew. One man stitches a patch over a hole in the birch-bark skin while

It now became necessary to consider how we should get on but the Canadian Voyageur soon finds a Remedy and our Men were immediately occupied in repairing the Hole. The Woods furnished the material. Bark from the Birch Tree Wattape from the Root of the Pine, Splints made from the Cedar Tree and the Crossbars. In the Evening all was ready to start in the Morning. DIARY OF NICHOLAS GARRY/ 1821

another melts pitch to waterproof the patch. The man in European dress may be an agent sent out from the fur traders' home office.

4/ The Ways of the Beaver

If one wants to have any idea of what the world once looked like, one must remember that it was inhabited by beavers.

LARS WILSSON/ *MY BEAVER COLONY*

Through the trees I see a distant gleam of silver. I am looking for a beaver pond, and I think I may have found it: where the silvery sheen beckons I know there is a shallow valley between two low hills. Cautiously, I move toward it. I have the impression that I am walking as soundlessly as any Indian, but I must be crashing through the brush as noisily as a moose on the trail of a mate—for suddenly I hear a sharp crack up ahead, followed by a splash. If beavers are there, I have certainly alarmed them, for the way they signal danger is to slap their broad, heavy tails against the water, producing a sound like a rifle shot. Then I hear other cracks and other splashes, and I hurry as best I can, convinced now that there is a whole family of beavers diving to safety as they hear me coming.

I emerge from the woods to see a small pond of such exquisite beauty and stillness that I feel like a wilderness transgressor just looking at it. Spruces border it, their reflections utterly clear in the perfect mirror image of the pond. The water at the edge is black, but so limpid that I can see the pebbles on the bottom. Only an occasional whisper of a breeze stirs the surface, crinkling it into little silvery ruffles. Not a sound is to be heard; even the birds are silent.

A few yards away, jutting into the water at the base of a dead tree that stretches bony fingers upward, is a big mound of sticks piled seemingly at random—a beaver lodge. Beyond it, at the lower end of the pond, is a beaver dam. As I walk along the water's edge to take a closer look at it, I notice the fresh prints of six-inch-long beaver paws in the mud at my feet. The dam, stretching perhaps 60 yards across the valley, is an old one, tightly compacted: grasses grow out of the tangle of sticks that an earlier beaver generation used in constructing it. At one side of the dam there is an obvious weak spot, and a trickle runs through it with tiny gurglings.

Tonight, after I am gone, the beavers will undoubtedly be out to work on the leak, small though it is. While some sort of spillway to relieve the pressure of backed-up water is as essential to a beaver dam as to a man-made one, this hydraulic necessity is of no moment whatever to beavers; so far as they are concerned, running water is there to be dammed. And a dam is vital to the beavers because of the pond it creates. This enables them to swim to trees on which they can feed after they exhaust the supply in the immediate vicinity. No less important, the pond provides near-perfect protection against enemies. The beavers usually build their lodge at the pond's edge with the entrance on the water side. As the water level rises, the lodge entrance is submerged. The beavers can make their way through this underwater tunnel; nonaquatic predators like the lynx and the wolf cannot. Security of food and shelter—this is what is represented in the laboriously fashioned dam and lodge I see before me at the small pond.

Beavers are among the most remarkable creatures of the north woods. They are unexpectedly big animals—the second largest rodents on earth, after the South American capybara, another aquatic animal that resembles a warthog. At maturity beavers range from 30 to 50 inches in length and weigh 30 to 40 pounds. Curiously, they never quite stop growing. Only death sets a limit to their size, usually after 10 or 11 years if they live in the wild or up to 19 years in captivity.

The hard-working habits of the beavers are of course legendary; they have become the very symbol of conscientious industry. This is because their dam-building and tree-felling activities seem extraordinary to man—as indeed they are. Once, on the Pigeon River in northern Minnesota, I saw a beaver-felled aspen that was at least 100 feet long. The tree was about 12 inches thick at the point where the beavers had neatly chewed it; it might have fed a beaver family for a month, what with all the succulent branches it carried. But unfortunately it had fallen away from the river instead of toward it, where the beavers could have chewed off parts to float them homeward. Moreover, the top of the

Standing amid its meal of water plants, a young bull moose with a spring growth of antlers enjoys the cool of a northern lake.

aspen was hung up in a couple of large black spruces, and the trunk leaned at about a 45-degree angle. I could see marks where the beavers had made their way a short distance up, but they are poor climbers, and the topmost branches remained beyond their grasp. The toppling of the tree was a formidable achievement, but it was labor lost.

To fell a tree, a beaver begins by biting lightly into the trunk all around; then, continuing to circle the tree, it bites off huge chips of the interior wood. To take these bigger bites it first locks its two upper incisors, which protrude an inch and a half past the lips, into the top of the chewing area. Below this point it uses its two lower incisors to cut a deep notch; then, with the upper incisors again, it finally pries off a large chip of wood—three, four or even five inches long. Thanks to its large loose lips—so loose that they can fold behind the teeth—the beaver is able to gnaw off such chips without getting a mouthful of wood. The same attribute permits it to gnaw wood under water without drowning.

A beaver's teeth are well suited to its needs. The front surfaces are coated with hard enamel, and as the uppers and lowers grind against each other and against the tough fibers of the animal's woody diet, they are constantly resharpened. Like the beaver itself, the teeth never stop growing. If the animal has little wood to chew—as is sometimes the case in captivity—the teeth will grow longer and longer, assuming weird curved shapes and eventually doubling back into the beaver's jaws until it dies a horribly painful death.

The speed with which a beaver works would put Paul Bunyan to shame. An inexperienced axman like myself might well have taken half a day to bring down the Pigeon River aspen; for the beaver it was probably no more than a couple of hours' work. To cut through the soft wood of five-inch-thick willow takes the beaver three minutes flat. Speed, however, is not always of the essence. There are known instances of beaver-felled trees that have obviously required longer effort; they measure five feet or more in diameter. Why a beaver should attack a tree that thick is a mystery; the animal cannot possibly haul so big a trunk to its pond, whether to store as food or to use for dam repairs. The supposition is that beavers chew trees simply because millennia of evolution have programed them to do so; but they have no way of knowing that one tree may be better situated than another for their particular needs, and they are certainly unable to make a tree fall in a specific direction. They simply chew around the trunk, usually a foot or a foot and a half up from the base, cutting a beautiful smooth notch. When the fibers start to give as the tree begins to sway, the beavers beat a hasty retreat, sometimes not hasty enough: occasionally, a beaver is found dead under the weight of a tree it has felled.

Despite such mishaps, the beaver's record of success is phenomenal, even more so in its activities as an engineer. The dams it constructs are veritable miracles of their kind. The naturalist and writer Sigurd Olson cites a beaver dam in Minnesota that was half a mile long and firm enough to hold a horse and wagon. There are many instances of dams a quarter of a mile or more long, and if they are old enough they will have compacted to the point where they will support almost any weight.

Beyond the fact that the dams are always built in running water, there seems to be no rhyme or reason whatever to their location. Sometimes the site selected is notably efficient for the beavers' needs, permitting the construction of a dam that backs up water for miles. In another site the dam appears to reflect nothing more than an excess of zeal; it is built across the widest part of a stream even though a narrower site may be just yards away.

In any case, the dams are usually elaborate affairs. The beavers start construction simply by poking sticks of all sizes—twigs, small branches, chewed saplings—into the mud in the stream bed of the projected site. Gradually they add sticks until a wooden barricade stretches from bank to bank. Since the sticks are not interwoven, water flows through at first. The beavers deal with this problem by periodically adding mud that they dredge up from the bottom and carry clasped to their chests with their front paws. They plaster the mud liberally wherever needed, patting it into place like children making mud pies. Later, drifting bits of debris pile up against the dam, further sealing it until the passage of water is almost completely blocked. But the beavers remain continuously at work, repairing the inevitable damage wrought by winter storms or spring floods. An old dam may reach as high as seven feet above the original water level, and it is impervious to just about anything except dynamite or a bulldozer.

In time its builders will abandon the dam and move elsewhere if they find that they have stripped the trees around the site of the food they need. But this takes considerable doing, for trees as far as 500 or 600 feet inshore from the beaver pond are within the animals' reach. This supplement to their food supply is made possible by a network of canals, each at least a foot and a half deep—the minimum depth a beaver requires to swim in. Some networks are very extensive, and on sloping ground they have what appear to be locks.

In a rare daytime foray, a black-masked raccoon prowls a pond for crayfish. It usually dunks its food repeatedly before devouring it.

Whether these canal systems are the product of the beavers' deliberate planning or merely the result of happenstance is a subject of dispute among scientists. In earlier times such canals were hailed as proof of the beavers' ingenuity, and there are naturalists today who support the view that the beavers intentionally dig the canals and construct the locks. Other naturalists are skeptical. What actually happens, they believe, is this: the wet coats of the beavers dribble water along the paths they follow to the food trees, the animals churn the mud as they travel back and forth, the branches and sticks they drag move the mud to the sides of the runway, and soon a "canal" forms as water from the pond moves into the trench. This school of thought does credit the beavers for the locks in the canals, holding them to be small dams the animals have built to halt the movement of water. But otherwise, according to these scientists, the seemingly sophisticated canal and lock system simply represents the beavers' response to the presence of running water and loose objects. In fact, they point out, not even running water need be present to stimulate beaver activity: in captivity, a beaver will build a "dam" in the corner of its cage from anything lying loose—a food dish, a stick, even a keeper's dropped glove.

There is little scientific quarrel about another of the beaver's engineering achievements—the beaver lodge. Although to the eye of an inexpert observer it may look like no more than a casual heap of sticks, stones and mud, it is actually a snug and spacious shelter in which a beaver family can be completely protected against the sub-zero cold of a north woods winter.

The homesite is always near the water, and usually right in it; occasionally the chosen location is around the base of a large tree, which serves to anchor the building materials—sticks as well as saplings that the beavers have felled and of which they have already eaten part. In the early stages of construction, these materials are simply piled up helter-skelter in a rough circle to form a large mound that is perhaps three to five feet high. As building progresses, mud is used to fill the chinks and crannies of the mound to shield the lodge against wind and weather —a particularly effective device in winter, when the mud freezes. Because beavers are air-breathing mammals, air must have a way of entering, and this is made possible by minute natural openings in the roof of the lodge.

Building the lodge high makes for an interior living room commodious enough to house the beaver, its mate, and as many as eight kits. But the construction job is not yet done. The beavers now dig two or more underwater tunnels into the lodge, each invisible to predators and deep enough to permit the animals to swim about freely beneath the ice in winter. One tunnel is a straight runway for ordinary passage into and out of the lodge and for the transport of food. The other tunnels wind and twist and serve as escape hatches. Another finishing touch is a narrow feeding platform just inside the entrance, about four inches above water level. Here the beaver eats its meal of sticks, tossing the peeled remnants back down the tunnel when it has finished. Perched on the platform, it can also shake the water out of its coat and groom itself a bit before entering the living room. A pair of toes on each hind foot is especially adapted for the grooming process: the toenails are split so that they form a comb, and as the beaver combs its dripping fur, it simultaneously removes parasites and redistributes the oils with which its hairs are impregnated, so that its coat is again waterproof.

Ordinarily the finished lodge will last the beavers indefinitely. Years after the first occupants have moved away to more plentiful feeding grounds, a new beaver family may settle in and find the lodge entirely habitable. On the other hand, it may fail even its original builders. Sometimes, if spring floods are excessive, the beavers' nearby dam may break, and the pond will drain out. If so, the entrance to the lodge will appear above water, destroying the beavers' security. But except for such eventualities the beaver lodge remains proof against predators. A lynx or a wolverine will claw in vain at its three- or four-foot-thick roof. At the mere threat of enemy invasion, the beavers can escape through one of their underwater tunnels. As winter tightens its grip and a thick sheet of ice prevents the animals from moving in and out of the water freely, another protective device comes into play. Knowing instinctively that they must provide against the cold season, the beavers spend late summer and fall accumulating sticks and branches and pushing this food supply into the mud close to the entrance tunnel. All they have to do in times of freeze or at other moments of adversity is to swim to the entrance and dip into the larder they have handily prepared. And so life in the lodge is generally peaceful and secure, with little pressure from the outside world.

A beaver family is close-knit; beavers mate for life. Their kits—one to eight in a litter—arrive in the spring after a gestation period of approximately 100 days. The kits are born fully furred and with their tiny eyes open; they weigh about a pound, and are about 15 inches long, including a three-and-a-half-inch tail. Their front teeth—orange in color like those of their parents and many other rodents—are amusingly bucked; with this toothy look and their silky brown fur, they are almost irresistibly appealing. Nor do they present problems of housebreaking; the kits excrete their body wastes in the water and keep themselves scrupulously clean. Instinctive swimmers, they may venture outside the lodge within a few days after birth, under watchful parental eyes. In these first days of their lives they are extremely playful. They sharpen their swimming skills with water-tag games, and it is obvious that they enjoy themselves hugely.

For six weeks or so the kits subsist on their mother's milk. After weaning, they feed on the bark of deciduous trees but, like their parents, they also vary this basic diet by eating underwater plants such as duckweed, eelgrass and water-lily shoots. They remain under the parental roof for two years. When the third annual litter arrives the firstborn leave their parents, either voluntarily or under the threat of forceful measures. By this time, in any case, they are old enough to fend for themselves and start their own families.

As they move toward adulthood, kits are especially vulnerable to predators. When one approaches, the entire beaver family will flee the lodge through one of the tunnels, dive below the pond surface, and wait in watery safety until the enemy gives up and goes away. During this emergency the large lungs of the beaver—the means nature has given it to adapt to underwater living—are a particular boon. With this powerful equipment a beaver can swim as far as a quarter of a mile without coming up to breathe; it can stay submerged for as long as 15 minutes. Usually that is long enough to discourage any predator watching for it to reemerge. In a pinch, it may sustain itself even longer, though if it goes too far beyond its limit it will, like any air-breathing mammal, inevitably drown. But a premature expression of curiosity may also prove its undoing. Sometimes a kit—or a parent—will poke its nose out of the water to see if the coast is clear and decide to emerge; at this point an alert carnivore can seize it.

A more efficient predator of the beaver has been man. During the heyday of the north woods fur trade he hunted and trapped the species almost to the point of extinction. Lately beavers have been making a comeback, although not in the numbers of the past. There have been relatively few human beings, it seems, who have viewed the beaver without the purpose of profit, and even fewer who have viewed it with gratitude. One such exception was John Colter, a guide on the 1804

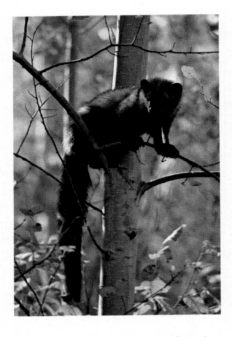

The fastest and strongest member of the weasel family, the fisher was misnamed by woodsmen who found it in fish-baited traps. But it seldom catches fish, preferring small forest creatures, which it stalks in trees (above) or outruns on the ground. It even preys on the porcupine, flipping it on its back to avoid the quills and attacking its tender underside. Vicious by nature, the fisher growls, hisses and snarls when provoked (right) and switches its foxlike tail menacingly.

Lewis and Clark Expedition, which blazed the first overland trail across North America. Colter reported on an experience he had at a beaver lodge in 1807, when he was on bad terms with a band of Blackfoot Indians. The Indians were intent on exterminating him and they all but had him when Colter suddenly saw a beaver lodge in a stream just ahead. Without pausing for an instant, he dived into the water, found the entrance tunnel to the lodge and swam right up into the beaver living room, where he remained hidden until the Indians had given up their search and departed.

Whatever the truth of Colter's story—and frontiersmen were given to fanciful accounts of their escapades—a more objective exploration of a beaver home was essayed in recent years by Leonard Lee Rue III, a naturalist and author. Rue was not being chased by Indians; he just wanted to see what the inside of a lodge looked like. It was a time of drought and the upper part of the entrance tunnel was above water. Rue stripped and slithered into the hole. He was half in the water and half out, in darkness that became absolute once he had rounded a corner about 12 feet in. At this point the roof of the tunnel sloped downward, and a few feet farther on he could no longer keep his head above water. Rue sensed a somewhat larger space ahead of him, and tried to get there by somersaulting forward. To his dismay he got stuck halfway through the maneuver, with his head under water and his feet jammed against the ceiling. "The only thought that flashed through my mind," he recalled later, "was, 'My God, what a place to get it.'" By superhuman effort he managed to reverse his somersault and get back upright again, after which he wriggled out of the tunnel backward. The experience left him wondering what might have happened had he made it all the way inside and discovered a beaver family in residence. Normally the beaver is a very inoffensive animal, but a family of six or seven, armed with long, sharp incisors, could very effectively rout an intruder in the homestead.

To beavers intrusion spells danger, and they are unable to distinguish between a predator, an innocuous investigator or, in my case that day at the little pond, a friendly writer simply anxious to catch a glimpse of them. I walked out on their dam, which I found to be only about six inches above the level of the water, yet very strong: the surface barely gave beneath my feet. No beavers were to be seen, and so I made my way back toward the beaver lodge. Wild flowers leaned out over the dark water, mirrored in the surface. A few ducks floated idly a little way offshore. As I walked along, wild rose bushes gently scraped at me, their dark red hips heavy and luscious against the deep green leaves.

There was nothing to be seen or heard at the lodge, and yet I had the feeling that beaver eyes were peering at me from somewhere. Finally I got close enough to see the food cache beneath the surface, between the lodge and the dam—an assortment of sticks and twigs, some with the green leaves still showing, piled at random like jackstraws. I waited there, listening hard for the sounds it is said one can sometimes hear coming up from the lodge below—the vague cries and whines of beaver kits or an occasional wail like that made by a hurt child. Hurt or frightened beavers sound almost like children, crying and moaning gently, while angry beavers hiss like cats. But I heard nothing. I was certain the animals were there somewhere, and just as certain that they felt safe from an interloper as unskilled as I.

I left my observation post as the sun was almost down. A few yards into the woods, in an aspen grove, I found a beaver feeding spot, littered with chips amidst the stumps of saplings and young trees. All the older trees had been felled. I sat against one for a little while to rest before going on and watched the darkening pond. The sun glared at me like an orange ball from the mirror of water.

Then, all of a sudden, I saw movement on the pond surface. The tip of a small "V" was cutting across it toward the lodge. I stared in fascination as the "V," a tiny black spot at its point, grew and spread, forming ripples touched with sunset red. Then I saw other "V"s being traced in the water, growing, spreading—and disappearing in the direction of the lodge. I walked away through the woods with a sense of exultation: for some strange reason I felt as though I had just joined the family.

Around the Beaver Pond

When beavers dam a stream to create a pond, they do so solely out of an instinct to establish a watery forest empire where they can obtain food from handy deciduous trees and at the same time ensure themselves a haven against their enemies —both animal and human.

But another effect of the beavers' building activities is to change their part of the forest profoundly, turning it into a rushy, sedge-grown, semi-aquatic world, part swamp, part lake. The shallow slow-moving waters attract a host of living things— from algae and plankton to fish and crustaceans. These creatures in turn have their own parasites and predators—from mosquitoes to hawks.

Yet even on bright spring days such as those that provided the pictures on the following pages, the beaver pond is deceptively calm. The few signs of life may include the soft splash of a frog leaping to safety from a patch of grass where a snake has surprised it, or a sudden ripple in mid-pond where a trout has struck at a caddis fly that was hovering too close to the water.

What cannot be seen are billions of microscopic plants and animals teeming beneath the placid surface —basic links in the food chain for nearly all other animal life in the pond. The beavers pay no more heed to them than to most of the other creatures with which they share their world. But in a sense the plankton are as dependent on the pond's beavers as the beavers, through the food chain, are ultimately dependent upon the plankton. For though the beavers may live in the pond for years, they will move on when they consume all the food trees in their reach. And just as they created ideal conditions for other forms of life while they remained, so they set the stage for the death of the pond when they leave. Their abandoned dam slowly disintegrates and releases its impounded water, the level of which drops. Supplies of oxygen and nutrients fall below the point where they can support aquatic creatures, large and small. Finally the pond dries out, to be succeeded by a stream-cut meadow.

The meadow in turn invites the encroachment of the forest. In a matter of decades, a new growth of willow and aspen will appear. And if the stream does not dry up, the combination of fresh food and running water will almost certainly attract a new family of beavers to begin the cycle once more.

A pond created by beavers near Ely, Minnesota, reflects their dome-shaped lodge in its waters. The spruces along the shore, though rejected by the beavers as a food source, will in time be felled by them for use in mending the lodge as well as their nearby dam.

A BEAVER FEEDING ON WILD CALLA

Signs of Spring in a Secluded Realm

As lord of the beaver pond, the animal that created it is unobtrusive in its dominance. Even after a long winter in the lodge on a diet of twigs and bark, it is rare to see a beaver feeding out in the open at pondside *(left)* on the calla leaves that surge up with the arrival of the first warm days of spring.

At this season, the most visible evidence of the beavers' renewed activity is the number of freshly felled trees around the fringes of the pond. Branches from such trees—which may be as big around as the two-foot aspen trunk pictured at right —provide both food and building materials. After the bark of the tree is stripped off and eaten, the twigs are used to fashion nests for the spring litters of kits. Longer branches go to build up the dam, thus helping to ensure a water level high enough to seal off the underwater tunnels that lead to the lodge. By this means predators are thwarted and protection is afforded the kits in their vulnerable first months.

Despite such precautions, danger still lurks for the beavers. Their worst threat is the otter, an agile swimmer that can negotiate the entrance to the lodge. At a hint that an otter—or any enemy—is in the vicinity, a beaver will slap the water noisily with its broad tail *(bottom right),* sounding an alarm to the other beavers and giving them enough time to dive to safety, hopefully far beyond the foe's reach.

A FRESHLY FELLED ASPEN

A BEAVER SLAPPING ITS TAIL IN ALARM

WILD CALLA

WATER LILY BLOSSOMS

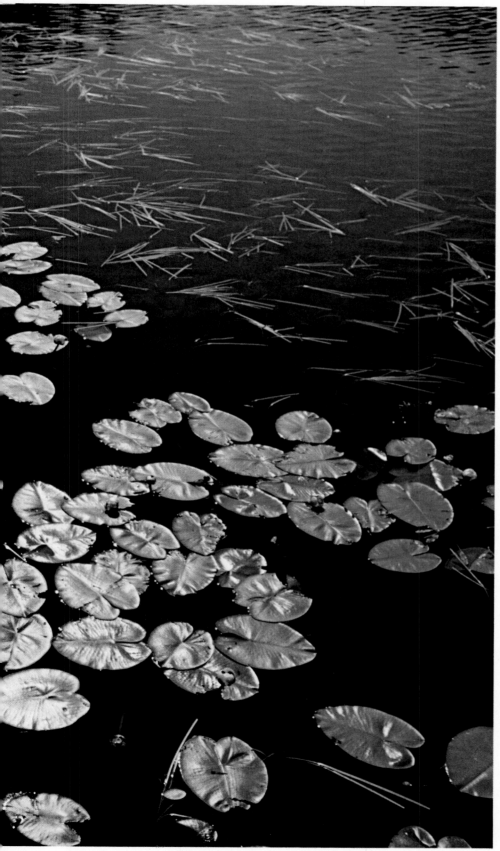

A Savage Struggle
Behind a Placid Façade

As the eye sees the plants and creatures of the beaver pond, and as the camera has caught them at left and on the following two pages, they reveal little of the intricate and frenzied pattern of competition and interdependence that rules their lives.

On a sunny spring day, the delicate yellow blossoms of the water lilies in the shallows charmingly offset the translucent blooms of the calla growing nearby. Yet gentle as they appear, these lilies can prove lethal to other plant life at the pond. Where they spread their wide pads, sunlight is cut off from other emergent plants, such as bulrushes, that are reaching for the surface. The first plants to capture a space survive on the pond's limited resources, but those that fail are numberless.

The struggle for survival is as fierce among the creatures of the pond. Seen in isolation, none of them seems either threatening or threatened. Yet the insect larvae that dot the surface of the water furnish the mink frog with the protein it needs to shake off the lethargy of winter. The frog in its turn may make a meal for the garter snake. Grass spiders, water scorpions, leeches and caddis flies all lead hazardous lives as potential consumers—and victims.

Transient though their individual lives may be, however, their kind will inevitably be back for another spring's struggle to share the pond with the lilies and the beavers, which endure from year to year.

GREATER YELLOW WATER LILY PADS AND FLOATING RUSH STALKS

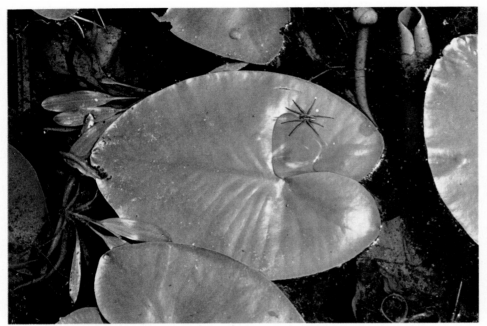

A GRASS SPIDER ON A LILY PAD

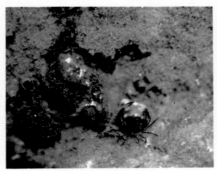

A LEECH ON A SANDY BOTTOM

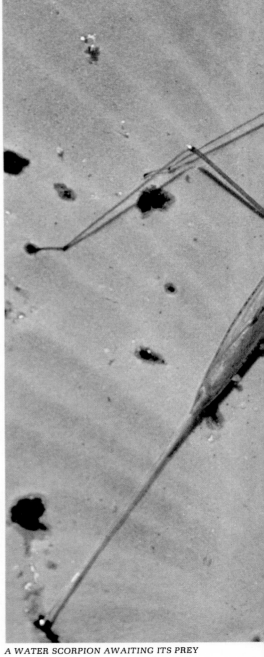

THE SHED PUPAL SKIN OF A CADDIS FLY

A WATER MITE AT PONDSIDE

A WATER SCORPION AWAITING ITS PREY

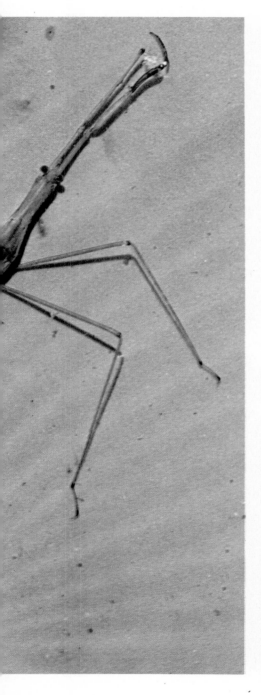

AN AROUSED EASTERN GARTER SNAKE

A MINK FROG ON THE MOVE

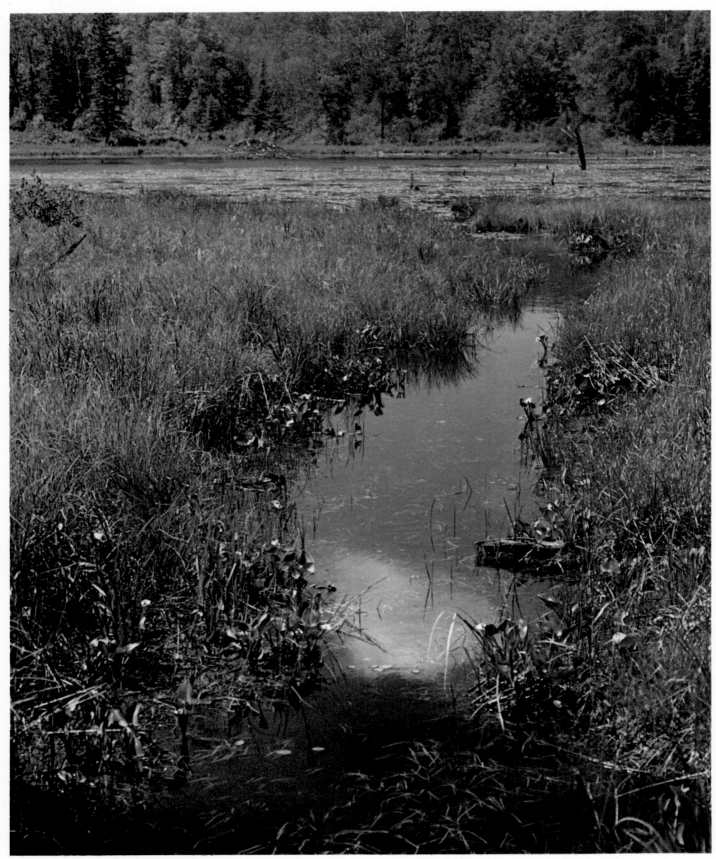

A BEAVER CANAL AT THE POND'S EDGE

A Perilous Venture in Search of Food

The need for food will often lure beavers beyond the pond's edges into the forest, a journey they make by means of a system of canals. Some zoologists believe that the beavers carve these channels inadvertently, simply by dragging their bodies through the swampy fringes of the pond; others credit the beavers with building these convenient passages purposely. In either case these channels serve the beavers well, providing access to fresh supplies of aspens and other food trees and serving as sluices in which bark and twigs can be floated from the outer boundaries of the pond to be eaten at the beaver lodge.

The tall grasses and thick woody growth through which the beaver canals wind support their own great variety of plants, insects and birds, some of which are pictured at right and overleaf. Birds build their nests in the area's grassy camouflage and their newly hatched young feed on a protein-rich diet of the pond's insects, which are attracted by the scents and spots of color provided by wild flowers. But hospitable as the boundary region may be for these species, for the beavers it represents a place of peril. Their clumsiness on land exposes them to such fearsome predators as the lynx, the wolf and the bear, and so they make their forays to new groves of food trees only briefly, scuttling back down their canals as fast as they can to the relative security of the pond.

A RUFFED GROUSE ON ITS NEST

A CLUTCH OF SPARROW EGGS

A CRANE FLY

SAWFLY LARVAE ON A HAZEL LEAF

A SWALLOWTAIL BUTTERFLY

BUNCHBERRY

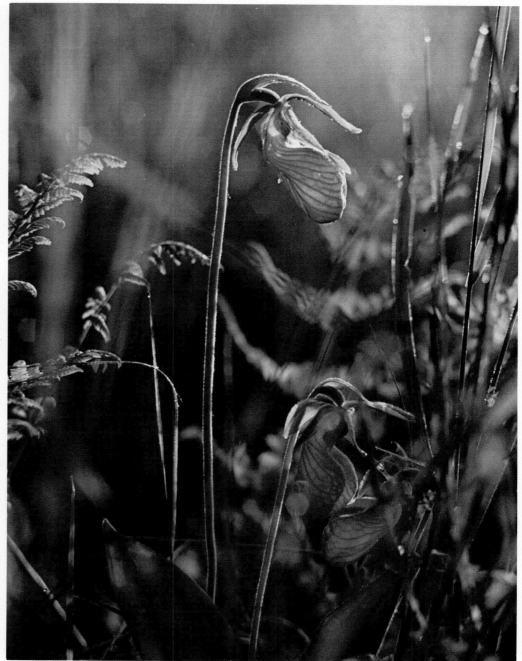

WILD IRIS

A MOCCASIN FLOWER

Confident as the relative safety of dusk descends, a beaver swims powerfully through the dim tranquillity of its pond's shallow waters.

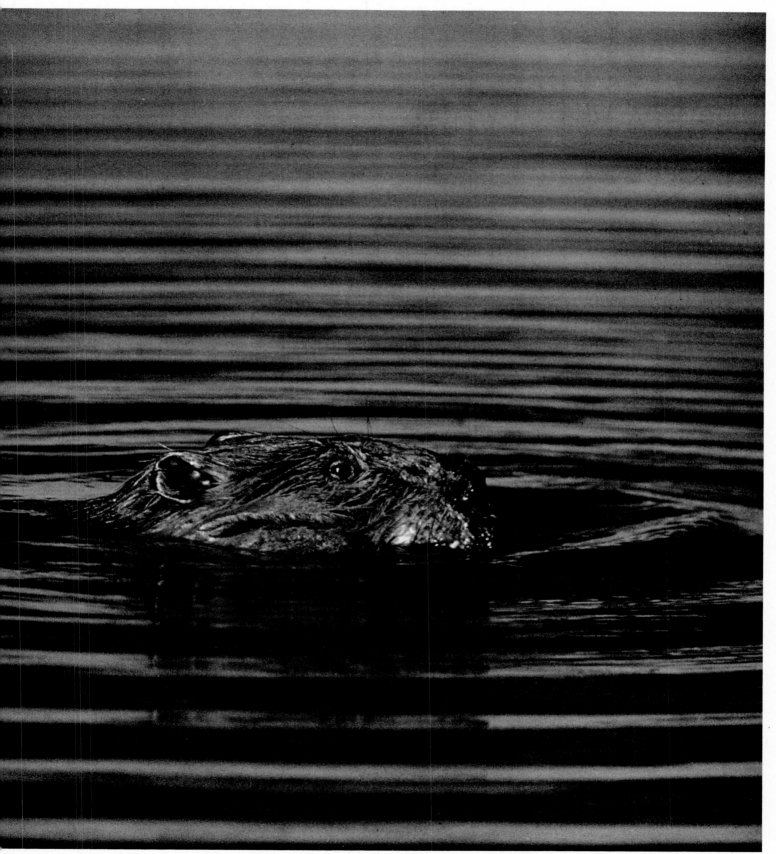

Most active after dark, the animal faces a busy night's work of building, repairing and foraging for food before returning to its home.

5/ In the Grip of Winter

*I love the deep silence of the midwinter woods. It
is a stillness you can rest your whole weight against...so
profound you are sure it will hold and last.*

FLORENCE PAGE JAQUES/ SNOWSHOE COUNTRY

At 1 o'clock on a frigid January morning I stand in the woods above
Snowbank Lake in northern Minnesota, trapped by the silvery magic of
a full moon. The snow lies deep all around me; one step from the trodden path and I would sink in it up to my hips. The cold—it is 40° below
zero—presses in upon my goose-down parka, my felt-lined boots and
my mittened hands. The moonlight, astonishingly brighter than any big-
city street lamp, reflects on the delicate traceries of birch and aspen
branches, the somber pyramids of spruce, the wayward bulk of jack
pines. The great shimmering curtains of the aurora borealis glide back
and forth like phosphorescent ripples in some mysterious sea. The stillness in the white-clad forest is eerie and beautiful.

In the north woods in winter, snow is the dominant force. It is an environment all its own, requiring radical adjustments in the lives of
people, of plants and, most notably, of animals. In all animals, the coming of snow triggers reactions that change their mode of living
fundamentally. Some flee southward; others retreat to the deathlike
sleep of hibernation; still others change their food, their habits, even
their coats and their coloring. In this sense, every forest animal leads a
double life—one of spring, summer and fall, and one completely fitted
for survival in the winter cold. What causes the changeover from one
life to the other is the onset of snow.

I had observed the first signs of winter four months before in late September, some 800 miles north of Snowbank Lake where the dwindling forest meets the tundra on the shores of Hudson Bay. Here the air at night had a razor's edge of chill to it, and in the early morning I could see slivers of ice forming along the mossy margins of a muskeg pool. At the tree line, the autumn colors were of an intensity I had never before encountered: flaming scarlets, vibrant greens, flaring yellows, rich browns. Wandering the area around the bay, I sensed an animal world in transition, poised to go southward or withdrawing into itself for protection against what was to come. The air was alive with birds preparing for long flights. Canada geese, snow geese and whistling swans swept in ragged files across the lowering sky. Within the fringes of the woods, small animals were storing their food for the long siege that would begin when the first fine flakes of snow blew in.

Now, in January, snow lies everywhere in the north country, blanketing both its woods and its waters. Those most typical of summer sounds—the roaring of a waterfall, the rustle and burble of rapids on some river—are silenced. Here and there, where a stream's current is too swift to freeze, a whisper of movement may be heard from the depths, but that is all. The lakes, too, have changed their character: frozen hard, they now serve the caribou as resting places when the bands come out of the forest to seek the safety of an open space; they serve the packs of hunting wolves as a highway in their pursuit of caribou and moose and deer. The surfaces of the larger lakes are covered with wind-driven snowdrifts two or three feet high, crested like waves.

In every respect, because of the peculiar qualities of the omnipresent snow, the north woods in winter are a different world. Those of us who live in more southerly latitudes may find this hard to grasp; we tend to consider snow only as a passing phenomenon, sometimes giving us pleasure on the ski slopes, sometimes creating a hazard on our highways. In the north, however, its impact is so profound and all-encompassing that the study of snow as a natural environment in its own right increasingly interests scientists. They are learning more and more about how it can harbor as well as bury life, and about the transformations it works not only on the animals of the north but on their forest habitats. They are finding that snow itself—far from being the simple white stuff laymen believe it to be—is of different kinds. They are even developing a vocabulary to distinguish these varieties, borrowing from the languages of the Eskimos, Indians and Lapps, the people of the snow who know it best.

One pioneer in the field, William O. Pruitt Jr., a zoologist at the University of Manitoba in Winnipeg, employs these varied terms for snow not only in his learned papers but in his everyday conversation. *Qali* —the "q" pronounced like a "k," but with a glottal stop—is the word he uses for "the snow that is on the trees." Snow on the ground is *api*. The bottom layer of *api* is *pukak,* which provides a winter habitat for a host of small mammals. *Upsik* is the hard-packed, wind-driven snow of open spaces such as large lakes and bogs. *Siqoqtoaq* is crusty snow whose top layer has thawed and refrozen, sometimes so hard that it gashes the legs of moose and caribou that break through it; the crust can immobilize these beasts as effectively as an electric fence.

Each of these varieties of snow has important consequences for the areas on which it lands. Take, to begin with, *qali*. To most of us, snow that clings to trees seems merely picturesque, conjuring up visions of Christmas cards. That it could work major changes in a forest seems inconceivable. Yet *qali* plays a vital role in the cycle of plant succession, based on the simple but significant fact that snow on the trees represents a burden that eventually may become too heavy for them to bear. Piling up in the wind-still forest, the load of snow finally will break off branches or the tops of trees. When this happens, a break is caused in the forest canopy—a "forest window"—which is the beginning of a forest glade.

The effect of such a window is to change the ecological balance of the area where it has appeared. Through the opening left by the broken tree more sunlight reaches the trees still standing, and in time the branches on the glade side proliferate. This shifts the center of gravity of the trees toward that side. Since more *qali* accumulates on the side with the heavier branches, the burden on the tree is increased until finally, some snowy winter, it falls. With each tree that is downed, the story is repeated until at last there is a glade of substantial size.

Dr. Pruitt has observed the creation of such glades in study plots in the subarctic forest over a period of several winters. He has also noted that when a forest window reaches a certain optimum size, the *qali* breakage stops. This, he believes, is probably due to the fact that at some point the wind becomes a factor: now that it can sweep into the glade, it shakes the *qali* from the branches so that heavy accumulations no longer occur.

Meanwhile, important things are happening on the ground. The dead and broken trees shed their needles onto the forest floor, choking out the moss carpet. Deciduous shrubs and trees invade and provide leaf litter, which decomposes into humus—and now a hospitable soil has been created for the seeds of spruce. In time, these take over, and finally a mature stage is reached in which the coniferous trees predominate, and the cycle of plant succession can start all over again.

Added to its key role in the life of the forest itself, *qali* plays a crucial part in denying or giving access to the food on which certain forest animals depend in winter. If the *qali* build-up is heavy, tree-living animals such as the red squirrel, the chickadee and the crossbill can no longer reach the spruce cones that are their chief source of sustenance. The squirrel must retreat to the forest floor and dig beneath the snow to seek out the cache of food it has prudently stored there, while the birds must fly off to windy hilltops where the trees are clear of *qali*.

Another familiar animal of the north woods, the snowshoe hare, finds *qali* a help rather than a hindrance. For its winter food it depends on the tender growing tips of birches and alders. These trees react differently to *qali* than do spruces. While spruces strive to remain upright and bear their growing burden of snow, the supple deciduous saplings bend with the weight, enough so that eventually their growing tips come within the hare's reach. Moreover, they provide the hare with a shelter beneath their bent branches during spells of extreme cold. In a classic example of interaction for mutual advantage, the trees provide food for the hare and the hare reciprocates by leaving pellets of fertilizing manure that help to nourish the trees when spring returns.

In some ways the hare has the best of all possible worlds in the winter woods. Its camouflage is superb; as autumn merges into winter, its brown and gray coat gradually turns snow white except for a thin rim of black at the tips of its ears. The snowshoelike hind feet that give it its name are another boon. When the hare can find no growing tips on which to browse and must subsist instead on the bark of birches or willows, it can rise on its hind legs and nibble all the bark within reach. At this point *api,* the snow that is on the ground, often comes to the hare's aid by building itself up another few inches and "elevating" the hare. Since the animal's snowshoe feet prevent it from sinking into the snow, it can now reach a higher level of the shrubs to browse on.

Api performs an even greater service in the case of small forest mammals and invertebrates. In the deep-freeze chill of northern winters its very presence makes it possible for animals of this size to survive, warmly sheltered beneath the snow itself.

To understand this, it is necessary to examine just what happens

112/

when snow falls on the forest floor. Where before there were mosses, lichens, a variety of small plants and considerable needle litter from countless conifers, there is now an insulating blanket that covers the forest floor entirely. Snow is one of the best insulators in the natural world: it is actually an emulsion of air and myriads of ice crystals. The basic shape of the crystals is a six-sided star, and because of this shape they are not able to nestle closely together. Hence the snow, when it first falls, is light and fluffy. As layer falls upon layer and as meteorological factors work upon each layer, the snow matures, and in the process many changes occur in its makeup.

The first change—the most important one to the small animals of the forest—occurs on the bottom layer, the *pukak*. The principal factors in this change are the warmth and moisture that flow from the earth. In the summer, the warmth and moisture simply radiate into the air and disappear; now they are trapped beneath an insulating cover. Warmed by the earth immediately below, the bottommost snowflakes begin to lose their water molecules, which flow off the attenuated rays of the six-sided crystals and gravitate to the colder crystals of the layers of snow above. Gradually, below the snow cover, an open space is formed that is interspersed with delicate crystals of ice, larger than most snowflakes and different in shape—hollow pyramids that hang together at their tips and form a delicate latticework of interlocking columns.

The *pukak* latticework may be as much as several inches thick —enough to offer a hospitable environment to any animal that may live there. Here, close to the base of the snow cover, the temperature is seldom more than a few degrees below freezing, no matter what it may be in the air above. In outside winter temperatures, small mammals such as the red-backed voles, the mice and the shrews would freeze solid in a short time; their bodies are simply not big enough to put out the metabolic heat needed to keep them alive. But in the relatively warm bioclimate of the *pukak* these animals can live, breed and reproduce. And they live in what any human being would regard as a veritable fairyland. The air is always warm and moist and still. The light that filters down through the snow cover is a pale bluish white. The only sounds are the scamper of tiny feet, the occasional tinkle of ice crystals falling from the roof of the *pukak* and the footfall of a predator stalking through the snowy woods above. Frigid winds may roar over the forest canopy; down in the *pukak* they are never heard. There is food aplenty, stored in the forest floor during the summer months, and there is seldom any need for vole or mouse or shrew to visit the world above.

Deep in Superior National Forest a September snowfall—first of the season—whitens a majestic stand of black spruces and balsam firs.

Once in a while, however, they must do so, and here again is demonstrated the balance of the winter world's ecology. In the layers of snow above the *pukak*, the maturation process is constantly taking place. There will probably come a time when a warm front, perhaps with a drizzle of rain, will briefly thaw the top layer of snow. When it refreezes, the snow cover will have a tough, crusty layer of ice on it, and this ice will impede the exchange of gases between the forest floor and the air above. Carbon dioxide formed by the decomposition of leaf litter will begin to fill the crystal halls of the *pukak* layer. When the carbon dioxide level becomes dangerous to animal life, the tiny mammals build ventilator shafts to the upper air. Through these, they can come up and catch a breath of cold fresh air—but this offers an opportunity to the little Richardson's owls that prey on them. The formation of a single layer of icy crust works to the benefit of these birds by bringing the mammals on which they feed within their reach.

The only other predator the voles and mice and shrews need fear in winter is the weasel, which is small enough to penetrate the *pukak* corridors. It may also happen that a fox passing by above will catch the scent of a *pukak* dweller and hear the tinkling sounds that signal a living presence below. The fox, jumping into the air with all four feet in order to come down hard, may try to break through the snow cover, but likely as not will only find itself in a snowslide from which it can get out only with difficulty. Or a moose or a deer may break through the cover with its hoofs, causing an avalanche of snow that blocks the *pukak* corridors. But soon the process by which the bottom layer of snow yields up its water molecules to the colder layers above will start anew, and in time the *pukak* will be as snug a habitat as before.

On frozen lakes and in open bogs, a different maturation process takes place, this one brought about by the wind. As the wind blows across the surface of the snow, it picks up the topmost layers of snowflakes and whirls them along, tumbling them over and over. As they move, the delicate snow crystals change their shape: the elongated rays break off, and instead of star-shaped snowflakes there are now countless millions of needle-shaped crystals of varying size. Blown along by the wind, these form side-by-side patterns, much denser than was possible with the star-shaped snowflakes. They then settle behind some protuberance—a crack in the ice of a lake, a rocky outcrop protruding through the peat of a bog—and drifts begin to form.

Where the wind blows fiercely and constantly, these drifts can be amazingly hard. Piled up five feet or more, the snow achieves a cementlike toughness: when struck with the flat side of an ax it rings like a bell. It is solid enough to support the foot of a man or the hoof of a caribou, and the running wolf leaves barely a footprint in it. This is the snow called *upsik;* when it melts and refreezes into a crust, as it may after a warm front occurs, the tough icy layer is known as *siqoqtoaq.*

Both *upsik* and *siqoqtoaq* crucially affect the winter lives of the large mammals of the north woods, sometimes for good and sometimes for ill. The moose, for example, has stiltlike legs that generally keep the bulk of the beast above the snow level. But it is too heavy to walk on snow. Deep snow spells trouble, and heavily crusted snow can cause the moose to crash through and become trapped. To avoid the *siqoqtoaq* it either migrates to a region of thinner snow cover or stays put in a "yard" where food is plentiful, moving about as little as possible. If it exhausts the bark and twig supply from the trees and bushes around it, its only recourse is to rise on its hind legs, bring the weight of its body to bear on the tree and bend it down to where the higher food can be reached. Sometimes this will break the tree: a moose's yard in springtime is usually marked by broken branches and tree trunks and—if the snow is crusty—by leg holes rimmed with blood.

In the case of caribou, *upsik* and *siqoqtoaq* serve a unique purpose that only now has begun to be understood, thanks to research by Dr. Pruitt. These two varieties of snow are determining factors in the strange, seemingly erratic migrations of the caribou—both the woodland kind, which is rare now except in the northern forests of Canada, and the Barren Ground kind, whose habitat is the tundra. Once the caribou wandered from tundra to forest and back again in herds numbering hundreds of thousands; today, though much depleted, the herds follow the same general course. What chiefly influences the direction of these wanderings are "fences" of snow.

Dr. Pruitt has observed the migrations of woodland and Barren Ground caribou over a period of years, charting their courses in northern Canada and Alaska. By collating these observations with his studies of different kinds of snow he has found the explanation for the animals' sometimes inexplicable behavior, notably in the springtime. Faced with warming trends that might be expected to lead them southward toward still warmer weather, the caribou head north instead—as if they wanted to stay in a wintry climate for as long as possible.

In summer caribou feed on sedges, grasses, lichens and the leaves of willows, birches and aspens. In winter, their deciduous food is gone; lichens and sedges are all they have left. To get these, they migrate to the woods and crop their nourishment from beneath the snow, pawing out "feeding craters" with their front hoofs so that they can reach down and get the food. Some types of snow—notably the light, fluffy *api* of the deep forest—make feeding easy for them. Other types, such as the dense, wind-blown *upsik* or the icy-crusted *siqoqtoaq* make feeding difficult. The caribou's migration thus has a double aim: to get them to places where they will find food, and to avoid areas where the snow makes feeding difficult or impossible.

Dr. Pruitt and his students have made extensive comparisons between areas where caribou are numerous in the winter and the types of snow found there. The animals are found in greatest numbers where the snow is light and fluffy enough to allow them to get at the food beneath it. They may not congregate in the same area the next winter; while the food may be there, it may be inaccessible because the snow is dense or crusty. Or there may be an area of ideal snow but with no food beneath it. And so the caribou push on, wandering aimlessly, it would seem, but actually guided by snow conditions—"fences"—fully as effective as any fence man might erect.

When the cold begins to loosen its grip and the sun sheds some warmth on the frozen earth, the topmost layer of snow may thaw and refreeze. The crust may get so thick that it cuts the skin of the caribou's legs, causing great discomfort and making feeding impossible. When this happens, the caribou herds head north—where the snow is likely to be still fluffy and soft. Sometimes, as spring nears, they find themselves squeezed between a fence of crusted snow that is building from the south and the open tundra, where the deciduous plants that are the caribou's source of summer foods are still deeply buried and weeks away from their budding stage. This is likely to be the most difficult time of the winter for the caribou. Such conditions, if they are unusually severe, may wipe out entire herds.

Whether they are wandering north or south, the caribou gravitate for their rest periods to the frozen lakes, whose surfaces offer a welcome refuge where they can chew their cuds and loaf and sleep. From these open spaces, too, they can spot the approach of their principal predators, wolves, and can bunch together and make a run for it with the easy, swinging trot that wolves cannot match. Wolves, for their part, must follow the caribou, a primary source of their food. Thus they, too, migrate from the tundra to the forest and establish hunting bases under the snow-laden branches of some large spruces. From here, in single

All hardy winterers in the north woods, the four species of owl at right endure the cold that other birds flee when fall arrives. The snowy owl, the great horned owl and the great gray owl are of imposing size, with wingspreads up to five feet. Richardson's owl is small as northern owls go, with a wingspread of only two feet. In all owls, the wing tips are notched, allowing the free and relatively noiseless passage of air when the birds are in flight. Thus their search for prey is virtually soundless —a prime advantage for a predator.

GREAT HORNED OWL

SNOWY OWL

RICHARDSON'S OWL

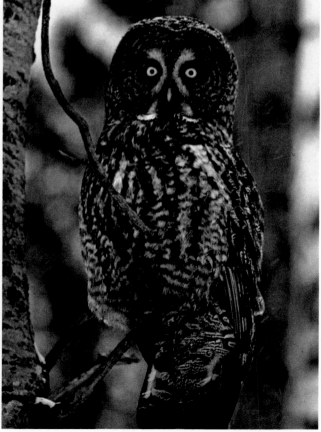

GREAT GRAY OWL

file, they make their trails through the snow, and almost always the trail will lead toward a lake, the caribou's resting ground.

Wolves do not stalk like lynx or mountain lion; theirs is a straightforward approach to the job of finding and killing their prey. If they spot a band of caribou on a lake, they will advance. When they are within sight, one of the caribou will spot them and instantly give the danger signal—a hind leg thrust out sideways. The band will spring up and, in a tight-packed bunch, explode into a clumsy gallop, then settle down to a distance-eating trot that they can keep up for hours.

When the caribou flee in this tight bunch, with no stragglers, the wolves seldom attempt to follow them, instinctively knowing they have little chance of success. But let one caribou hesitate or falter, out of indecision or because of age, sickness or injury, and that caribou is doomed. Such behavior instantly triggers in the wolves the stimulus to kill. There is an old saying that "a wolf can catch any animal it chases" and it is true to the extent that the wolf usually will chase only an animal it thinks it can bring down. Wolves may test many caribou or moose before they encounter one whose flight reactions deviate from normal. Even so, unwittingly but with superlative efficiency, the wolves cull the caribou herds. By killing the weak and infirm, they help to maintain the quality of the herd and establish the fitness of the caribou to survive. This is a form of interdependence between wolf and caribou no less essential to both than that between spruce and *qali*.

There is only one documented instance in North America of a wolf attacking a human being; in 1942 a man riding a handcar at 10 miles per hour on a Canadian railroad track was set upon by a wolf, which hung on for almost half an hour before three other men succeeded in killing it. The animal, however, was believed to be rabid. The rarity of such encounters is probably due to the fact that men do not arouse the pattern of stimuli required to cause a wolf to attack. My friend Sigurd Olson, who has lived in the north woods all his life and studied wolves in their own habitat, tells a revealing story in this regard. Skiing down a frozen river one moonlit winter night, homeward bound, he became aware that a pack of wolves was following him. He had no gun, just a hunting knife, but he also had his own knowledge that wolves do not normally attack man. "I had written a book that said that," he recalls with a grin, "but I didn't know if the wolves had read the book, so after a while I got a little nervous, realizing that what I had said about them not attacking man might well be put to the test that very night.

"Up ahead of me, reaching more than halfway across the width of the river, I could see a rocky point. I would have to ski past this, through a narrow gap, and I began to think to myself: 'If they're going to jump me, that's where they'll do it.' I slipped my knife out of its sheath and held it in my hand, thinking that if I had to die, I would at least try to take one of them with me. Now I could see the point clearly in the moonlight, and as I looked, two big wolves came out of the woods, trotted down the rocks and onto the river, and sat there facing me. There were probably more out of sight in the woods.

"I was perhaps within 30 yards of them, and not knowing what to do, I stopped. I stood still and looked at them, and they sat still and looked at me. Predator and prey, I thought; but which is the predator now, the wolf or me? The knife didn't feel very reassuring, but it was all I had, and I clutched it as though it were King Arthur's sword.

"We stood there and looked at each other for what seemed a long time. The eyes of the wolves—I could see them clearly in the bright moonlight—were unwavering. I didn't move a muscle. Finally the larger of the two wolves got up, shook himself, trotted up the rocks and disappeared into the woods. The other got up and followed him. And I skied down the river until I got to my cabin, which I then entered, being extremely careful to bar the door."

Was it the fact that wolves fear man that kept those two unmoving in the snow? Or was it the fact that the man who faced them stood unmoving, betraying no sign of fear or weakness that might provoke the deadly stimuli that unleash the wolves' attack behavior?

I was thinking about this as I stood above Snowbank Lake on that January night when suddenly I heard the most beautiful, the most chilling sound to be heard in the north: the calling of the wolves. It came from over the ridge, probably from a half mile away. There was a single voice at first, rising in a long wail. Another joined it, then another; then the wolves were singing in harmony, as I had heard they often did.

The sounds aroused some stimuli in me. The hackles on my neck rose, a cold vise gripped my heart and traveled toward my stomach. Without thinking, I found myself walking quietly through the snow to the warmth and safety of my cabin, where I, too, barred the door.

The Bane and Blessing of Snow

Winter in the north country is a time of harshness. Daylight is brief and warmth scant. A none-too-easy environment for animal life even in milder seasons, the north woods can support even less life at this time of year, as temperatures plummet and the snow builds up, denying access to customary food supplies. Yet, paradoxically, it is the snow itself that helps forest animals survive the winter. For those that can exploit its characteristics, the snow can be friend and protector.

Certain animals—called chionophobes by zoologists, after the Greek word *chion,* or snow—cannot tolerate the snow at all. Notable among these are ground-feeding birds and waterfowl; they take to the air and migrate south at the first chill hint of winter. Other animals, called chioneuphores—moose, wolves and such small mammals as voles and shrews—make adjustments to the snow. Wolves hole up in soft snowbanks to conserve body heat when the cold becomes unbearable. Moose easily wade through depths of soft snow on their long stiltlike legs. Voles and shrews take up winter residence in crystal tunnels beneath the snow cover: the layers of snow serve as an insulator, trapping the ground heat below the bottommost layer and affording these tiny creatures a haven in which the winter temperature seldom drops below 20° F.

A third group of animals, called chionophiles, or snow lovers, have made actual physical adaptations to winter conditions. The broad feet of the predaceous lynx enable it to move surely and swiftly across even relatively fluffy snow cover. The snowshoe hare possesses similarly well-adapted hind feet and also another advantage: a pelt that turns from brown to a camouflaging white as the snow cover builds up.

But despite adjustments and adaptations, life remains difficult in the north woods winter. Carbon dioxide occasionally fills the tunnels of the voles and shrews and compels them to surface for a breath of fresh air and to create ventilation shafts, risking a pounce by a hungry fox or owl. Moose and deer may break through crustings of snow, gashing their legs. The snowshoe hare, emerging incautiously from a warm burrow in the snow, may be seized by an alert lynx. Across the snow the trails and tracks of predators and potential prey attest to their perilous search for the food that will enable them to survive to see another spring.

A nameless island in Manitoba's frozen Wallace Lake catches the sunrise. At this hour, nothing is astir, but there is life here; nocturnal animals have retired to await another darkness, while diurnal creatures do not yet dare venture into early-morning gloom.

The Deceptive Delicacy of a Snow Called Qali

No variety of snow has a greater impact on the vegetation of the north woods—and thus on the animals that eat the vegetation—than *qali,* the feathery sort that rests on trees. Its branch-borne beauty is deceptive; *qali* is a potent force in the forest.

Its accumulating weight causes such deciduous trees as birches, willows and aspens to bend temporarily to the ground, making their tender tops accessible to the snowshoe hare. Conifers will not bend under the *qali* unless they are very young, as in the case of the jack pine at left. Instead, such trees will break, beginning with their branches. As more trees topple, a "window" is created in the forest canopy, allowing sunlight to reach the forest floor for extended periods and nurture the deciduous trees on which deer and moose can browse in later winters.

The big deterrent to *qali* is a strong wind; sweeping in through the forest window, it will prevent the pileup of snow. Snow and wind often work together to aid animal life, notably on the open spaces of lakes. The wind compacts fluffy accumulations, in time hardening the lake surface enough to permit such creatures as foxes, wolves and caribou to use it as a winter highway.

A young jack pine bends nearly to the breaking point under its burden of seemingly lightweight qali. Both the cold and the absence of wind have permitted the build-up of snow in these woods near Wallace Lake, Manitoba.

A solid highway for animal travel will eventually evolve atop these fragile hoarfrost crystals on the surface of Wallace Lake. The crystals provide a base for falling snow, which strong winds then gradually pack hard enough to support even heavyweight animals.

A Final Round in the Contest to Survive

That most avid of snow lovers, the hare, is not without its troubles in the winter woods. If the snow is even slightly compacted, the snowshoelike feet of the animal will carry it lickety-split across the surface in search of food trees. But if the snow is too soft the hare will sink in. To cope with this dilemma it hops up and down to carve out trails to feeding areas. Each new soft snowfall, however, obliterates the trails—and the hares, impelled by a special behavioral adaptation, must leave their snow caves, usually at night, to reestablish their routes.

But as they engage in their frenzied hopping dance, the enemy lurks. The lean, furry lynx—the only cat adapted to the northern forest—can lie patiently and motionlessly for hours beneath the branches of a tree before striking. Like the hare, the lynx has enlarged feet that enable it to move swiftly in quest of a meal. The hare is the mainstay of the lynx diet, and when it ventures from the cover of the trees into an open area its movements can give it away despite its camouflaging white winter pelt. Quickly and efficiently, the lynx pounces. The cat's long teeth crunch through the hare's skull into its brain, killing it instantly.

A snowshoe hare falls victim to its classic enemy, the lynx. As a rule the foes are equally agile, but this hare made the error of taking a short cut through a clearing where its movements could be easily spotted.

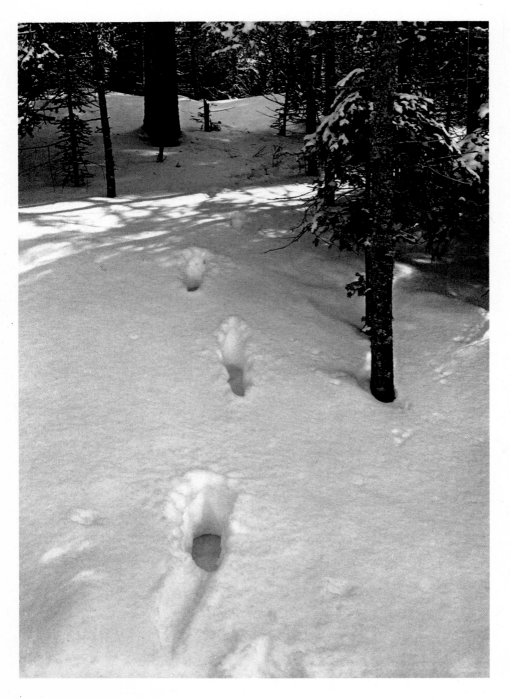

Signs of Life in a White Wilderness

The visitor to the wintered-in north woods seldom sees any of its animal inhabitants, but evidence of their presence abounds in the form of trails and tracks. In the snow, prints of hoofs, paws, wings, tiny feet and whole small bodies not only betray the passage of living creatures but reveal much about their habits and methods of survival in winter.

Of the five animals whose telltale marks are shown here, the moose leaves the most heavily indented imprint, thanks to the long legs that enable it to move through deep snow; the tracks often lead to a "yard" of relatively open space where it can find succulent deciduous trees, the moose's favored winter eating. The grouse, burrowing beneath soft snow for nighttime warmth, makes both entrance and exit holes recognizable by an observer. The flying squirrel, seeking another food tree, drops from a branch in a glide, then skitters swiftly to its goal. Wolves, though they hunt in single file, following a trailbreaker through the snow, leave rough circles of paw prints after a kill or a period of play. The hopping feet of hares form a fairly direct line—a trail that takes them straight to their treasured sources of food.

Deep hoofprints in the snow cover mark the long strides of a moose, which has made its ponderous way toward an area in the forest in which it can stay put for a period and browse on branches conveniently nearby.

Entrance (left) and exit holes denote a grouse's nighttime burrow.

A glide and a skitter of paws reveal a flying squirrel's route.

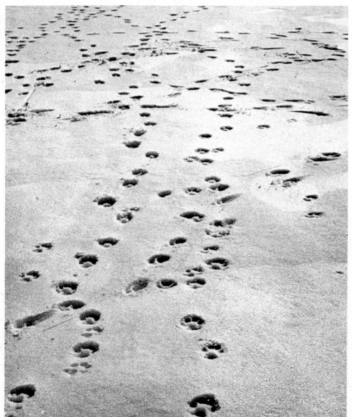

Tracks made by a milling wolf pack cover the surface of a lake.

The hopping of snowshoe hares marks a clear trail to their food.

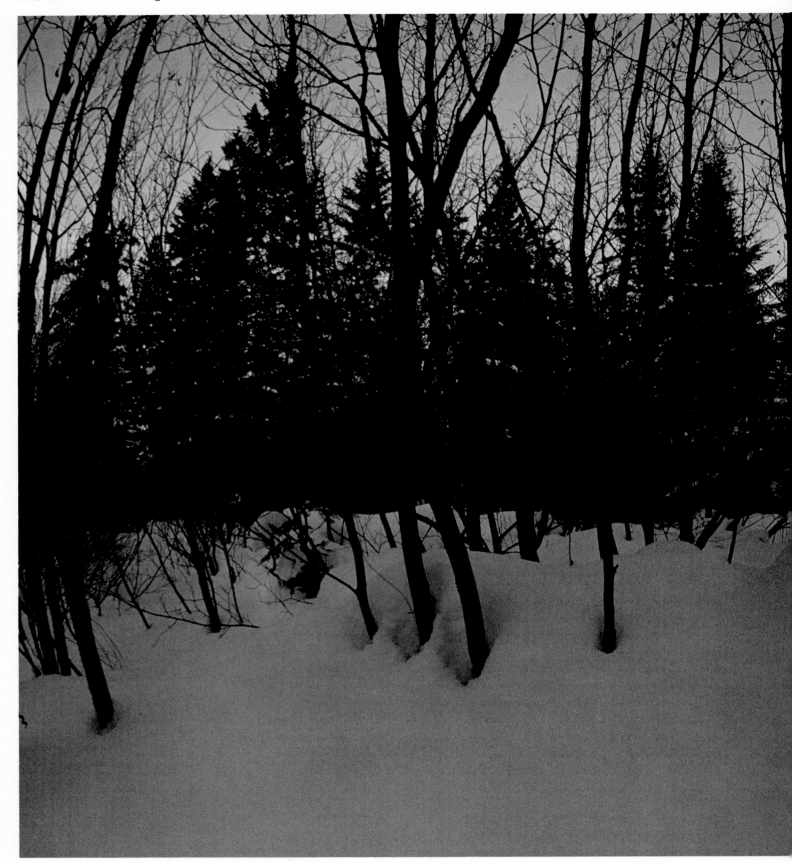

The snow pyramid at right covers a mound-shaped beaver lodge and provides an added shield against predators. But the residents face a

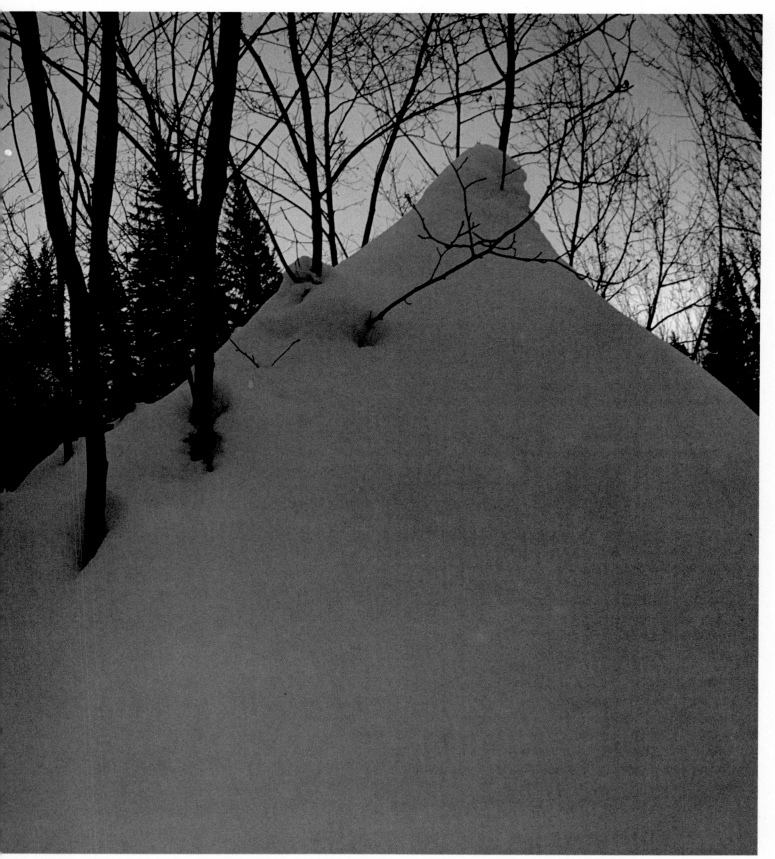

new danger in winter. In the dark lodge, the beavers are deprived of the light they require for growth, they lose weight, and some die.

6/ The Burgeoning Spring

*In the spring I look at the forest floor with
excitement, no matter how dead it may appear to be, for
life is there, waiting for the signal to emerge.*

SIGURD F. OLSON AND LES BLACKLOCK/ *THE HIDDEN FOREST*

Spring, that sweet and so often romanticized season, is not universally beloved in the north woods. For man, it is a time of paralysis; he cannot move for days on end. The ice that provided his winter highways is crumbling now; he can neither walk on it nor go through it in a boat or a canoe, and not even bush planes can land on the thawing lakes and rivers. In the woods, trails turn treacherous with mud and flood. Slush is everywhere as the snow blanket slowly turns to water—slush that loads up skis and snowshoes until the feet can scarcely lift them. "Summer is too short to go through the hell of breakup," say oldtimers, and they retire to their cabins, stoke the fire and sit out the melting season.

Sometimes it seems as if the whole world is liquefying. Or so it seemed to me one early April day as I tried to cross a woodland meadow in Manitoba. The muddy path had been churned deep, probably by a moose's hoofs, and I decided to bypass it, stepping along the sides where patches of dried grass promised a more secure surface. But suddenly one of my feet sank to the ankle in ooze. I put my weight on the other foot in order to pull free; instead I pitched forward. Somehow I managed to get my mired foot loose and took another step; this time I sank halfway to my knees in a gluelike mass of mud. And so it went for a while. Each time I put weight on one foot it would sink deeper as I tried to pull the other free. I could go neither forward nor back, and I did not see how I was going to extricate myself.

In such situations large hoofed animals like moose or deer may become hopelessly ensnared as though in quicksand. But whereas in quicksand they soon sink out of sight and suffocate, in mud they linger, immobilized. They may linger thus for days until at last either a predator gets them or they succumb slowly and agonizingly to starvation. Whether this has ever happened to men I cannot say, but after this experience I could believe that it has. I finally rescued myself by the most delicate maneuvers, shifting tentatively and precariously from one foot to another. I had only about five yards to go in the mud, but it took me all of half an hour, panting with exertion, to reach firm ground.

With the perils of spring thaw, however, there is also a joyous sense of renewal in the north woods. Amid the muted colors of winter splashes of brightness appear. Rocky slopes glint with the blue of harebells; silvery-pearl pussy willows dot the banks of a creek; flaming stems of red-osier dogwood fringe a marsh. The transformation is swift. Exploring along the Peace River in April 1793, Alexander Mackenzie noted in his journal: "The change in the appearance of nature was as sudden as it was pleasing, for a few days only were passed away since the ground was covered with snow."

Sounds as well as sights announce the northern spring. The long silence of winter is broken by the songs of whitethroats and marsh wrens, the drumming of grouse, the whine of mosquitoes aroused from their hibernation. But these are gentle sounds compared to two others that above all epitomize the start of the new season: the wondrous noises of ice breaking and of water unleashed.

Within the woods the dominant theme is that of ceaselessly running water, myriads of separate trickles joined in crescendo. As a balmy sun sheds hour after hour of radiation upon the earth, and a wind bereft of its knife-thrust chill blows warmth in among the trees, the ice crystals of the snow cover begin to melt. For the small animals that the snow has housed, this is a time of trial. Their tunnels collapse; more vulnerable now to predators, they are forced to seek haven elsewhere. But the melting goes on relentlessly. Water seeps everywhere, dripping from ridges and rocks, dribbling along the ground. Soon minuscule streams merge, perhaps on a trail, forming rivulets and finding a firm direction. For the final goal of all the running water in the woods is the river or the lake; the rivulets take their last leap from some stony ledge or mossy bank and unite with the waters of the shore.

In the open spaces of lakes and rivers the drama of spring thaw is more easily observed, an ear-piercing, eye-filling spectacle that once witnessed is not soon forgotten. On the rivers it begins when the ice, weakened at the edges by the flow of meltwater from the banks, reacts by trying to move. In the attempt it creaks and strains and shudders, but at first fails. With more runoff it succeeds—cracking, then ponderously getting under way. Now comes one of nature's most awesome feats: in a succession of thunderous roars, the ice mass breaks apart. Great blocks of it float free. The tumult continues as they collide and shatter and overturn. At sharp bends or other obstacles in the river, the floes pile up, screeching as if thousands of automobile brakes had been suddenly and simultaneously applied. If the ice jam lasts more than momentarily, the liberated water behind it will back up and overflow the banks, flooding the woods. Even after the jam loosens the danger to the shore persists. Blocks of ice careening downstream will slam into the banks, uprooting trees, shearing off branches, stripping bark.

Sometimes the ice breaks up suddenly enough to strand an animal that happens to have been using it as a highway; spectators along the shore have been treated to the poignant sight of a startled doe passing by on a floe. In a big river, where the current is strong and swift, the breakup may come and go quickly, and people place bets on the length of time the process will take. I was at Fort McMurray, in Alberta, when the Athabasca River went out in 31 dramatic minutes, making one lucky bettor richer by about $800. During the previous week, under a genial sun, I had watched the daily widening of the telltale dark band along the shoreline, evidence that the edges of the river ice were giving way to reveal the water below. On the day of the breakup itself, the powerful current swept the whole ice pan downstream in a rushing maelstrom of splintering floes that broke into ever smaller fragments until at last the river flowed high and clear. The effect was that of a violent explosion, except that it took place not in an instant, but in the relative slow motion of half an hour.

On the lakes, the breakup is more deliberate. One might describe it less as a melting than as a rotting process. At the same time the ice melts from top to bottom it also deteriorates internally. In the first stage, the warming sun melts the snow that covers the lake ice. When the cover has turned to slush and water, the sun attacks the ice directly. At this point the crystals of the ice undergo a structural change. Throughout the winter they had been lined up side by side to form a substance almost as hard as iron; now they form honeycombs through which water drips constantly as the surface ice melts. Gradually the walls of the honeycombs become thinner and thinner. The ice loses all of its inner strength; it may still be six feet thick, but the weight of a foot, either of man or animal, will collapse it: one breaks through the surface as readily as through a snowdrift, with a sound like the shattering of a thousand delicate glass goblets.

There are two phenomena particularly associated with the breakup of ice on a lake. One is the lifting of the ice; it may rise as much as a foot, sometimes with dramatic suddenness. The lifting occurs because the lake basin at the time of thaw fills up with meltwater above as well as underneath the ice cover; thus the ice, which is lighter than water, is subjected to two pressures—the weight of the water on top of it, offset by the stronger thrust of rising water from below. When the shore ice has melted and the ice can float free, it responds to these stresses by rising in turn. If the shore ice has melted unevenly, one side of the ice cover may tilt upward while the other side is still held prisoner.

In time, of course, the warmth of spring prevails. At the final stage of melting, the lake ice darkens, and when this happens, oldtimers in the north know that its hours are numbered. The crystal honeycombs have now become so thin and fragile that they no longer reflect the light from above, but reveal the darkness of the depths below. If the wind is calm, the ice will simply vanish, sometimes in a single day: suddenly it is no longer there. If the wind is strong, it will drive the ice along until it piles up on shore; there, under the battering of the waves, it breaks into glittering shards that ride up on the rocks with a melodious tinkling. If the day is sunny, the beauty of the scene is breathtaking: the shoreline flashes as if blanketed with diamonds, and out on the lake whitecaps form like victorious battalions, inexorably advancing against those last remaining battlements of winter's cold.

With the final disappearance of the ice, one of the most fascinating of all the rites of spring takes place in the lake. Over the winter, the ice cover effectively seals off the vital exchange of gases between the air above it and the water below. Whatever oxygen remains in the water continues to be consumed both by fish and by decomposing plants and other underwater life. By winter's end, in relatively shallow lakes, this oxygen supply may be critically depleted, sometimes so much that the fish suffocate. The arrival of spring is a literal lifesaver, for with the departure of the ice the waters of the lake completely intermix, redistributing oxygen to every level of the lake.

This renewal occurs as the sun heats the surface water, newly freed of ice. When the surface temperature reaches 39.2° F., the temperature of all the waters in the lake, from top to bottom, becomes relatively uniform, and so does their density. It is this stability of conditions that permits a thorough mixing—when a strong spring wind blows. The turbulence produced by the wind makes the whole lake circulate, bringing oxygen-poor bottom water to the surface to become recharged.

In every level of the lake the fish begin to stir; they may now swim anywhere. Trout that in summer will elude the fisherman by staying in the colder depths appear among reefs along the shore, their golden brown and reddish hues plainly visible. I remember one spring evening at Wallace Lake in southern Manitoba. At one end of the lake there is a spot where the water is very low; it is a marsh full of water plants, cattails and reeds. Fish were not on my mind at the moment; I was hoping for the sight of moose. As I waited quietly in the woods just above the shore, I heard repeated splashes from the direction of the lake, as though fish were leaping out of the water after flies. But there were no flies around, nor any mosquitoes. I tiptoed to the shoreline and looked for the source of the sound. Finally I saw it: a spot where a tiny swirl of water was being pushed up like a bubble. The bubble was moving swiftly through the reeds, directly toward me. As I watched in rapt attention, I saw the slender, sinuous form of the fish beneath the bubble —a northern pike nearly three feet long. It swam almost to the shore, nosing about on the bottom as it did so; then with a swish of its tail—a movement that broke the surface of the water and produced a splash like those I had been hearing—it turned and was gone.

I have enjoyed the onset of spring in many places in the north country, but nowhere have I found it more interesting or more unusual than in the northeastern corner of the province of Alberta. Here, in a sprawling arc around Fort Chipewyan on Lake Athabasca is an enormous delta formed by three great rivers: the Athabasca, the Peace and the Slave. The Peace-Athabasca Delta, as it is known, covers 1.5 million acres, and has long been a wildlife paradise. But there are other wildlife paradises on earth; what makes the delta unique is that in the spring, as well as in the fall, it serves as a funnel for every one of the four major flyways on the North American continent. Vast numbers of water birds that migrate back north after wintering in warmer climes travel through the delta; more than a million of them stay to breed or stop to feed there en route to Arctic nesting areas.

The singular hospitality afforded them by the delta was made possible by two natural factors: the flatness of the terrain and a remarkable

Doomed to disappear in a day or so, the disintegrating ice on Minnesota's Hegman Lake lies under attack by the warmth of a mid-May sun.

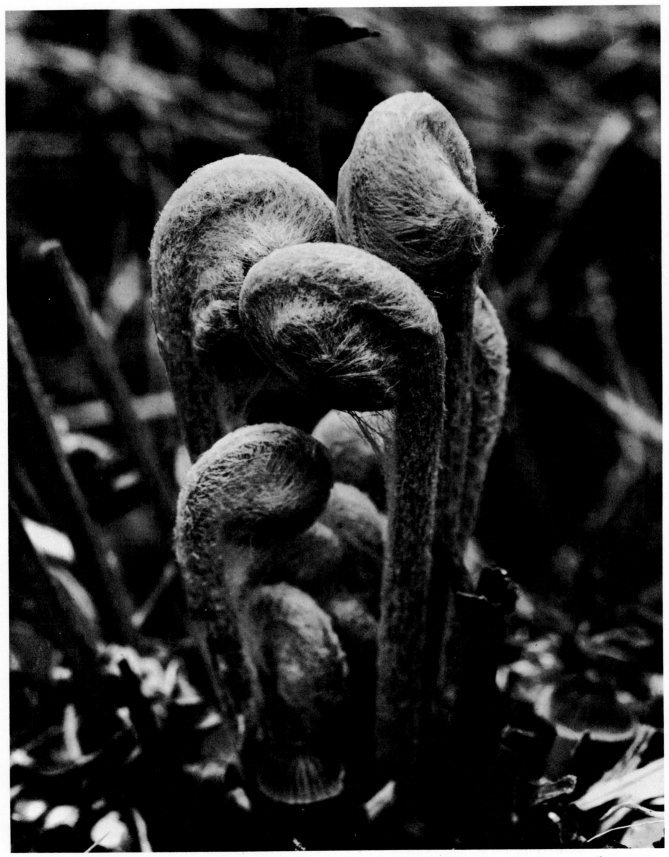

Harbingers of spring in the woods, cinnamon ferns will gradually unfurl their fiddleheads as the warmer weather persists.

hydrological relationship between Lake Athabasca and the rivers, particularly the Peace River. One might expect this river, powerfully coursing eastward from the Canadian Rockies, to drain into the lake; instead, the lake drains into the river. The Peace veers north just before it gets to Lake Athabasca, and the lake water flows into it through two of the delta's many complex channels. Each year in late spring, swelled by the runoff from melting snow and ice, the Peace would rise to a level higher than that of the lake. For about 20 days the lake water would be blocked from its normal flow. As a result it would back off into the delta, flooding a considerable portion. The floodwaters would distribute their heavy load of silt and rich nutrients throughout the marshes and numerous smaller lakes and streams of the area, revivifying its plant life and providing a gigantic larder for its animal life.

Moose that have fed on low willows and other shrubs through the winter now have a plethora of water plants on which to browse. In the ponds and marshes muskrats gorge on eelgrass and bulrushes; these semiaquatic rodents have outstripped the beaver population that attracted the fur traders of old. Seen grazing in meadows, on the grasslike sedges, is the oddest looking of all the delta animals: the wood buffalo, a subspecies of bison. This shaggy, humpbacked brown beast is the largest land mammal in North America, weighing as much as 2,400 pounds. There are about 14,000 of them in the area; unlike moose, which tend to be loners, the wood buffalo move in herds. One bright morning, from the vantage point of a light plane, a friend and I spotted a small band of them peacefully chomping in an upland. As we flew lower for a better look, the animals stampeded into clumsy flight.

They had nothing to fear from us, nor for that matter from any man. As long ago as the 1890s, at a time when extinction threatened the entire bison species, the Canadian government legislated to preserve it from human predators, assigning the Northwest Mounted Police to enforce the law. The establishment of Wood Buffalo National Park, which includes most of the delta and millions more acres as well, has strengthened the protection. Throughout the park its namesakes roam wild, while overhead the bald eagle and the peregrine falcon, almost extinct elsewhere on the continent, wheel free and secure.

But it is the waterfowl, in all their astonishing variety and numbers, that make the delta most memorable for me. They appear as the weather warms, and the delta comes alive with their calling. Whistling swans fly arrow straight across the marshes, their wings beating the air with rushing sounds. All sorts of ducks and great gaggles of Canada geese occupy the mud flats and shallows. There are herons and grebes, loons and cormorants. There are several species of gulls, and terns that defend their nests with the aggressive skill of fighter pilots. The delta is also the breeding ground of the white pelican, the strange bird "whose beak holds more than its belly can," as the old limerick has it.

If the delta visitor is really lucky, he may catch a glimpse of that rarest of birds, *Grus americana*, the whooping crane. It is a very impressive sight, standing nearly five feet tall on long, spindly legs, its plumage a pure glossy white except for black face markings and wing tips. Its yellow eyes, gazing fiercely upon the world, seem to reflect the spirit of the bird. It is rigid in its ways, unable to adapt to any habitat other than marshland that is rich in vegetation and small marine creatures.

Several decades ago—when fewer than 50 whooping cranes remained anywhere—efforts were begun to save the species. With the establishment of the Aransas National Wildlife Refuge, near Corpus Christi, Texas, a suitable wintering ground was provided, and it is there that the birds return, year after year, to spend the cold months. But where they went in the warm months was, until 1955, a mystery. Ornithologists knew only that the place was somewhere in northern Canada. Then aerial surveys helped them pinpoint the exact location—Wood Buffalo National Park. The whooping cranes that arrive there in the wake of winter, coming down out of the sky with the wild bugling call that gave them their name, still number no more than about 80, but at least the species maintains its precarious hold on life.

To see the delta in springtime is to witness one of the world's great natural sights. As far as the eye can probe there stretches a flat and sparkling mosaic that is both water and earth. The broad courses of the Peace, the Slave and the Athabasca seem laid out by design, so regular are their borders where the ice has melted. The innumerable smaller rivers and creeks meander every which way across an immense sprawl of brown, yellow and pale green bog and marshland. From above it all looks like a child's painting, a mass of mixed colors through which experimenting fingers have drawn large swirls and looping trails.

But now a shadow hangs over the delta, a shadow cast by the works of man. In 1967, some 650 water miles to the west of the delta, a hydroelectric dam went into operation on the Peace River. The harnessing of the river was probably inevitable; there is a continually growing demand for power all across Canada, and especially along its west coast.

Although biologists warned that construction of the W. A. C. Bennett Dam could lessen the annual rising of the Peace River, officialdom did not listen. Within a year it became evident that the effects on the delta's ecology might be very serious indeed. Coincidental with the fill-up of the dam's reservoir, the water levels in the delta have decreased. The dam appears to have controlled the flow of the Peace River in such a way that it can no longer be depended upon to rise high enough to make the waters of Lake Athabasca back up and flood the delta.

The results are beginning to be evident in creek and marsh and lake. Along the shorelines of lakes, now exposed to the sun, new growths of dwarf willows are pushing out the reeds and bulrushes on which muskrats thrive, and if this continues the muskrats will have to forsake these places. The wood buffalo, too, are affected. The sedges on which they feed depend for their growth on the annual flooding.

What will happen to the waterfowl no one yet knows. The ducks, for example, find in the delta an ideal habitat of small ponds and marshes; a single duck family stakes a claim to one of these as its own territory. Many ponds and marshes are beginning to dry out, and if this process persists the ducks may be forced to share the waters of the larger lakes with others of their kind and with other bird species as well.

Ecological disaster could engulf the delta, but there is strong hope that it will not. An intergovernmental task force, using small planes, helicopters, all-terrain vehicles and boats, has intensively studied every phase of the delta's complex life system. Hydrologists, biologists, chemists, geographers, engineers and experts in other scientific disciplines have taken part and have proposed solutions for the delta's dilemma. One solution would be the construction of small temporary rock-fill dams at strategic sites, impounding waters from the secondary rivers to raise the level of the lakes and ponds; one such dam, which would help to flood about 60 per cent of the delta, has already been completed. The birds and muskrats and moose and wood buffalo may yet take heart at the coming of spring; given the will and effort of man, their wilderness can be preserved.

NATURE WALK /To the Virgin Forest

PHOTOGRAPHS BY ROBERT WALCH

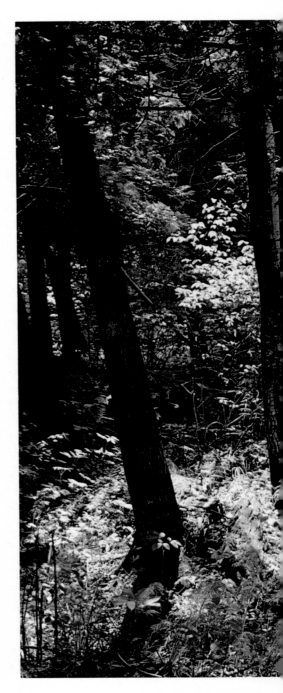

About a mile out in a lake in north-ern Minnesota, not far from the U.S.-Canadian border, lies a wooded island of priceless value to any lov-er or student of nature: an enclave that has never been logged. Today it is in private hands, secured against the power saw—a living testament to the virgin forests that flourished elsewhere in the north woods before they were wiped out by logging op-erations almost a century ago.

I have made the trip to the island a number of times. It is not a jour-ney of heroic proportions: from the starting point back of the lakeshore it covers a distance of only a few miles. It is, rather, a journey of the senses, in which increasing familiar-ity with a well-loved scene heightens my awareness of new growth and seasonal change. And so it was on my most recent venture to the island on a day in early June.

My trip began, as do so many north woods journeys, in a clearing at the end of a road. Before me stretched a typical second-growth forest, successor to a mature forest that the woodsman's axes and saws had destroyed. Slender birches inter-spersed jack pines and aspens. Two of the taller aspens had obviously provided a home for generations of

pileated woodpeckers; toward the tops of the trunks I could count at least six nest holes that these large red-crested birds had bored. As I ap-proached, two infant woodpeckers poked their heads querulously out of one nest, presumably awaiting their parents' return with food.

At the edge of the clearing, an old portage trail begins, leading through the forest toward the lake. Slowly I followed the ribbon of trodden earth through underbrush turned lush and green by spring rains. In the soggy ground along the trail, marsh mari-golds bloomed in golden bursts, the waxy flowers shining in the bright early-morning sunlight that gleamed through the trees. The air was musty with damp and the smell of decay-ing wood. Black water stood in several deep holes left where trees had been uprooted in previous sea-sons. Afloat in these pools, mosquito larvae were slowly hatching in the warm sun; already the air was abuzz with mature insects seeking their warm-blooded prey. A few mosqui-toes were caught in tree-hung spider webs that shimmered and swayed in the gentle breeze.

Thrusting up in thickets along the trail, most of the coiled new leaves of lady ferns had unwound, opening like fingers on an unclenched fist. Bracken ferns, and the horsetail so often found in their company, also flourished in profusion, prospering in the soggy ground.

As I approached the lake, the trail sloped downward and the ground became even soggier. The surround-

MARSH MARIGOLD

ing underbrush was thick with willow saplings, supple shoots of alder and an occasional young red maple. Off to my left I glimpsed the ruin of a small cabin, built about 70 years before by a Finnish woodsman. The roof had fallen in long ago, but the walls were still sound. I briefly detoured, as on past trips, to take a look inside. Rocks were piled atop the rusty remains of a crude stove. In the old days, when the stones were thoroughly heated and water was thrown on them, crackling steam would explode through the

room, well serving the Finn's purpose: he built the place as a sauna.

Near the lake, the old portage trail swerves onto a long neck of land that terminates in a rocky point. I walked past stands of white cedar, whose green fronds are a common sight in this area. White cedar has one of the

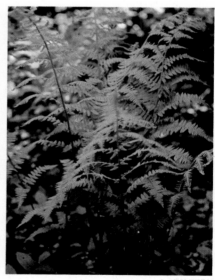

LADY FERN

most wonderful fragrances that I know of. Crush a frond in your hand and the pleasantly pungent scent will pour off it, the very essence of the north woods. Sprouting in and around the clumps of cedar were the broad-leaved shoots of another aromatic plant—sarsaparilla.

Along this stretch of trail I always seek out a tiny flower called goldthread. It has a starlike white blossom about half the size of a dime, so frail that it seems ethereal. But the real beauty lies concealed beneath the soil—in the root. Dig with a care-

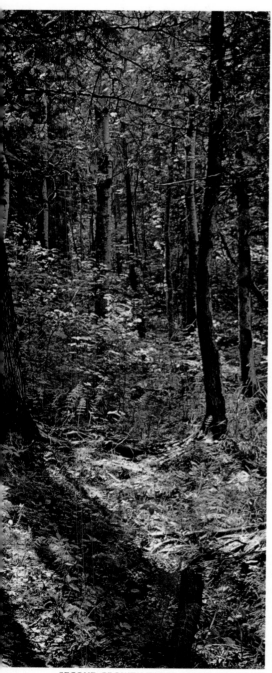

SECOND-GROWTH TREES ALONG THE TRAIL

ful index finger and you will find the root to be a long, slender thread the color of purest gold.

The Lakeshore Landscape

A few steps more brought me to the lakeshore and a curving beach of fine white sand. Just back of it, in the higher, drier places among the pines, berry bushes grew in dense tangles: bearberry, blueberry, woodland strawberry, and my own favorite, Juneberry, which—despite its name—bursts into masses of dainty white blooms even before the month arrives, a true indicator of spring. All these varieties of berry were flourishing only a few yards from the water, in close company with juniper and dwarf dogwood.

Almost at the water's edge I found a memento of the voyageur era: a clump of fleur-de-lis, once the royal flower of France. A member of the iris family, the fleur-de-lis is believed to have been brought to the north woods by early French explorers and voyageurs who planted it nostalgically at their forts and, occasionally, at campsites along the canoe route to the north—from which places water-carried seeds spread the species. At this time of year the fleur-de-lis was still putting its strength into green growth, but in a few weeks its graceful yellow flowers would shoot up between swordlike leaves.

Also near the water's edge I came upon the lilylike leaves and yellow blossoms of clintonia. Only a few of the fragile flowers had opened to the warm sun; most of the buds were still tightly closed. Hugging the

JUNEBERRY IN BLOOM

A FLEUR-DE-LIS IN EARLY GROWTH

FLOWERING CLINTONIA

ground, shaded by the clintonia's leaves, violets bloomed, tiny splashes of cool color that accentuated the rippling waters of the lakeside.

I continued along the shore toward the rocky point, skirting a large boulder deposited by a retreating glacier millennia ago. From trees closely overhanging the shoreline came the bell-like song of a thrush, accompanied by the excited whistling of nuthatches and chickadees.

A branch of one overhanging tree brushed my head. As I looked up, I saw a typical jack-pine cone, warped and twisted, sealed as tightly as if it had been dipped in hot wax. Someday, heat from a forest fire will break the seal and release the seeds, which will then be carried away by the wind. Somewhere a number of them will take root, for the jack pine is one of the toughest, most persistent species in the forest.

At the rocky point, a classic north woods vista unfolded. The rock is greenstone, and part of the Canadian Shield. The ledge shelved gently down into blue waters shaded by a lone red pine, standing sentinel-like at the tip of the point, its branches framing the view of a small wooded island just offshore.

The Trip across the Lake

From the point, I set out by canoe, paddling slowly to savor every inch of the way. Beyond the small island that I saw from the point two outcroppings of rock rise up from the lake. One has a bit of soil from which a solitary jack pine grows. I knew that in years past gulls had nested on this islet, and sure enough, a pair

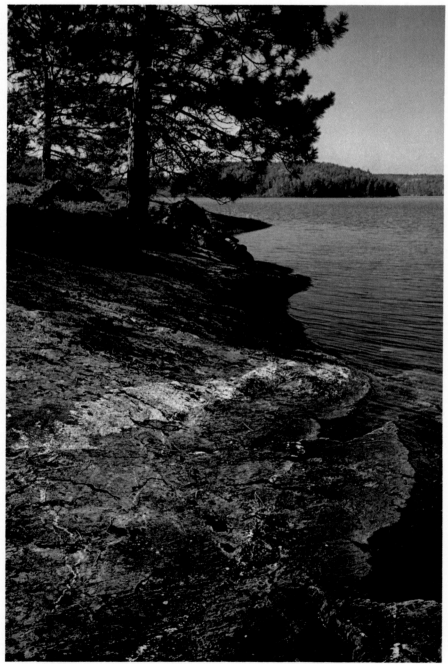

A RED PINE AT THE ROCKY POINT

APPROACHING THE ISLAND

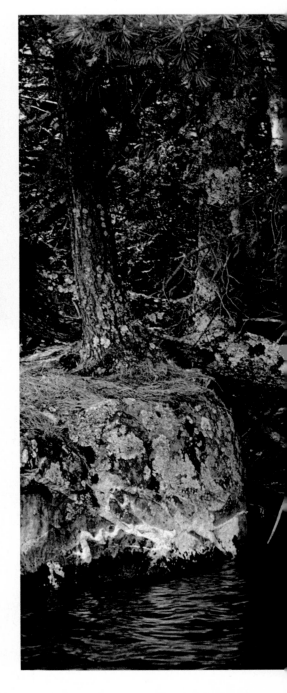

of them had taken up residence once again. As my canoe approached, the female gull winged up from a crude nest of twigs on the rocks; she was quickly joined by her mate. Frantically defending their unborn brood, they screamed and wheeled and made shallow dives at me with their beaks wide open. I paddled close enough to make an inspection of the nest. In it were three large eggs, dark brown with olive-green splotches. The mottled coloration very much resembled army camouflage.

I swung around another islet and saw my ultimate destination in the middle of the lake: a small island of about 10 acres, a place of cliffs and steep hills crowned by dense forest. Even from afar it has the look of virgin wilderness, a pristine little world

of its own that has managed to elude human alterations of any sort.

As I glided toward the shore, a rabble of ravens greeted me. Never had I seen so many of these big birds gathered together in one spot. From their perch on a large rock, two croaking ravens took to the air on sooty-black wings, and two more flapped awkwardly inland, out of sight beyond the trees. On the shore, other ravens assaulted the air with a raucous din, fighting furiously—perhaps battling for the remains of some small creature I could not see.

Slowly I circumnavigated the island. With every stroke, my paddle stirred golden pollen dropped by the minuscule blooming flowers of red, white and jack pines. At each breath of wind, more and more of the mar-

BRANCHES GROWING SKYWARD FROM A FALLEN WHITE PINE

velous powder was wafted into the water. The lake was streaked with gold, and where wavelets lapped against the shore they built up the pollen accumulation, forming what one poetic observer of the north woods has called "the yellow band of spring." Some of the pollen had blown across from the mainland; most of it, however, was drifting down into the water from the magnificent forest on the island.

A Study in Majesty

The forest reached all the way to the rocky shoreline, a dark green study in majesty, its maturity proclaimed by the size of spruces and red and white pines. There were no birches or aspens or other second-growth trees to be seen among these

A POLLEN-STREAKED LEDGE

splendid conifers. Along the shore several pines lay prostrate, felled by winter winds. One invincible white pine had somehow maintained a roothold in the earth after it fell, and though its top was buried in water, two branches had grown straight up from its recumbent trunk, seemingly trees on their own.

I put ashore on a low pollen-

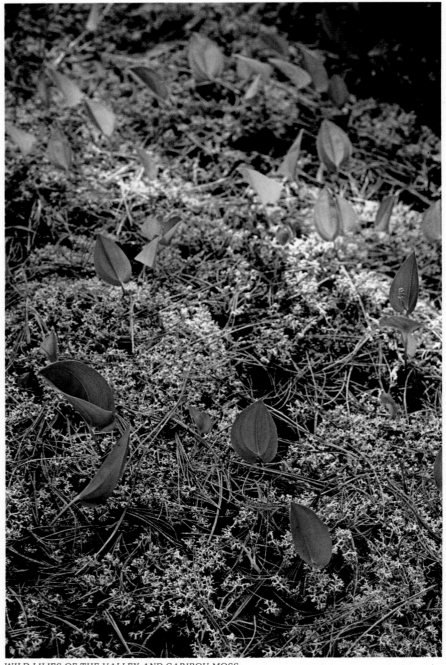

WILD LILIES OF THE VALLEY AND CARIBOU MOSS

streaked ledge and started walking inland through the trees. The terrain rises and falls sharply; rugged hillocks are interspersed with swampy hollows and defiles lined by rocks. Most of the island is covered with a thick carpet of pine needles. Slick and springy underfoot, the litter is replenished year after year by a steady rain of needles from the canopy of pines above.

The Domain of the Lichen

Here and there, under my feet, the layer of needles was penetrated by pearly gray-green caribou moss, a lichen named for the animals that make it a major staple of their diet elsewhere in the north country. In turn heavy clumps of this delicate plant were themselves pierced by several other plants, among them the wild lily of the valley, with its bright, heart-shaped green leaves and thick clusters of fine white blossoms.

Lichens grew on weathered rock in many varieties and endless profusion. There was one striking variety with slender gray branches and a bright red top, for which it has come to be called British soldier. There were crustose lichens, with almost microscopically small orange and yellow tissues, that from a distance look like splatters of bright paint. There were foliose lichens of a more somber hue, notably the *tripe de roche,* or rock tripe, once favored by the voyageurs as an addition to their soup. Its broad lobes form a blackish-brown coating on the sunny sides of boulders and cliffs; in hot weather they are as brittle as scorched paper but, boiled in a pea-

A SPARROW'S FALLEN NEST

CUSHION MOSS

soup mixture, they become moist and fleshy and quite tasty.

At the center of the island is its highest hill. Its granite flanks are covered with small, compact rock ferns, which grow even on vertical surfaces as if they had been glued in place. Higher up, the thick needle litter takes over again, sifting from the pines and softening the prostrate outlines of long-dead trees that are slowly crumbling back into the earth from which they came. The slippery needles made walking difficult, but after considerable exertion I finally reached the top of the hill. At my feet I found a cushion moss growing in a luscious, bright green mound as large as a Victorian lady's bonnet and just as decorative. Under a nearby spruce lay a fallen bird's nest —probably once the home of a tree sparrow. It had been blown down by the wind, which moans almost constantly through the trees.

Even though it was a warm June day, the wind swept across the lake in gusts that I could almost see. Each gust enveloped the island with an enormous sigh and started the tree branches swaying in a massive but graceful ballet. Even the island's tallest tree, a huge white pine near the crest of the hill on which I stood, joined in the dance.

I walked a few yards downslope to inspect the giant pine. Its trunk was about 30 inches in diameter at the base, with bark that was thick and rough but pleasing to the touch. The tree soared straight upward to a height of at least 100 feet. It must have been very old, possibly 300 years or more, to have grown so tall; I like to imagine that it stood watch on this hill as successive companies of voyageurs paddled by. The trees standing in the immediate neighborhood seemed to have retreated from it in awe. Actually, of course, its tre-

BRITISH SOLDIER AMID PINE-NEEDLE LITTER

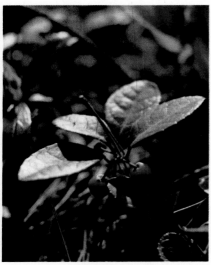
WINTERGREEN WITH BERRIES

mendous shadow has inhibited the growth of young trees nearby, leaving a circle of open ground.

Continuing down the far side of the hill, I saw my first sure sign of a land animal: a furtive flurry of movement in the undergrowth near a red-berried wintergreen bush. I stopped short and stood stock-still, hoping that the creature would show itself. Most likely it was a red squirrel, which abounds in these parts. But it could have been a chipmunk or even a chocolate-brown mink; I had seen both animals here on previous visits. I had also heard that an occasional black bear or a porcupine swims over from the mainland.

For two or three minutes I waited motionless. But I was out of luck: the creature had managed to make good its escape.

Time and the Forest

Suddenly it was late afternoon, time to head back to my canoe. The hike was a slow one, down one needle-slippery slope and up another, through or around swampy bottoms and rocky crevices. The wind was abating. A hush settled on the long colonnades of trees, and the ruffled waters of the lake became still. Above, the ravens circled silently on motionless wings.

As I walked back through this rare and beautiful wilderness, I thought, as always, about its future. By a rough count of tree types from place to place, I knew that the pines outnumber the spruces. This means that the forest, though clearly a mature one, has not yet reached its climax stage, when spruces and balsams

RANKS OF RED PINE

will be the predominant species. In the north woods, a forest requires about 400 to 500 years of natural development to grow from pioneer seedlings to this final stage. The island wilderness will therefore reach its climax roughly two centuries from now—barring some disaster. If such does occur, it is likely to be natural rather than man-made, for private ownership of the island guards against the latter eventuality.

But nature does, of course, inflict calamities of its own. The trunks of several trees on the island bear deep scars left by a forest fire that was probably touched off by a bolt of lightning. The fire undoubtedly took place many decades ago, for the scars were nearly overgrown. The bark of these trees had formed thick scablike tissue around the wounds, but the burned wood nevertheless showed through. A certain amount of burning, however, furthers the natural development of a forest, enriching the soil with the materials it burns, and so I hoped that the island would get its fair share of fires—all of them minor. Certainly they would pose less of a threat than the dangers once posed by loggers.

Why this virgin wilderness escaped the depredations suffered by the forests around the lakeshore became apparent as I turned to take a last look from my canoe, now halfway across to the mainland. The island's good fortune has been its size and location. To the loggers, it was simply too small and too inconvenient to reach from the mainland for them to discern any profit in it.

FIRE SCAR ON A PINE TRUNK

As I paddled shoreward at dusk, I conjured up an imaginary scene that might have preserved the island's virgin state the last time it was in jeopardy. In my mind's eye, the foreman of a logging camp landed on the island to make a survey. Up through the gorges he struggled, up the hills and cliffs, stumbling and cursing as he slipped and slid among the deadfall. The tall pines looked down upon his efforts with disdain. Finally, sweating profusely, the foreman returned to the spot where he had tied up his canoe. With a last frustrated glance at the forest, he muttered: "It's more trouble than it's worth." And so he paddled off across the lake, leaving untouched a living museum of north woods history.

THE ISLAND AT DUSK

7/ The Canoe—Key to the Lakes

These days in a canoe brought life in a new dimension. In calm waters we seemed to glide somewhere between earth and heaven, silently and gently.

WILLIAM O. DOUGLAS/ *MY WILDERNESS: EAST TO KATAHDIN*

"It was not so much a test of the body," once wrote Eric Sevareid, the Columbia Broadcasting System commentator, recalling a canoe trip he had made into the north woods as a young man. "The body takes care of itself at that age. It was a test of will and imagination, and they too, at 17, have a power and potency which rarely again return to a man in like measure. I would follow shock troops across a hundred invasion beaches before I would repeat that youthful experience of the rivers." Sevareid's journey, undertaken with a high-school friend, Walter C. Port, who later became a department head in a Minnesota store, remains a classic in the modern annals of the north woods. In 1930 the two boys, in a secondhand canvas canoe, traveled from Minneapolis to Hudson Bay, a distance of 2,250 miles via interconnecting rivers, lakes and portages. Their voyage cannot be precisely duplicated today; there are too many man-made dams, locks and other obstacles between Minneapolis and York Factory, the old Hudson's Bay Company trading post where the two young voyageurs eventually ended their journey. But parts of the trip can be retraced; I have traveled over sections of their route myself, as have a number of others, and can testify that canoe trips in the north woods remain a challenge to the will and a delight to the imagination. They are, in fact, balm to the soul.

North woods canoeing today is far easier than it was when Sevareid and Port made their voyage. To be sure, there have been trips in modern times on which people have died—because of mistaken judgment in rapids, because they ran behind schedule and were caught by hunger or cold, or because they got lost in unknown territory. But this does not happen often: the experienced modern voyageur carries detailed maps and an accurate compass, and packs freeze-dried food that can be stowed so compactly that he can take a supply for many weeks. And with lightweight nylon tents and goose-down sleeping bags, he can survive almost any kind of weather.

Sevareid and Port almost did not make it. They lacked reliable maps, they had few woodland or navigational skills, and they were facing what was—and remains—some of the wildest country on the North American continent. The last 500 miles of their journey, from Norway House at the northern end of Lake Winnipeg to York Factory, led through territory virtually uninhabited and largely unexplored. Some Cree Indians lived there, visited on occasion by missionaries and Royal Canadian Mounties, and a few trappers still staked out their lines in the territory. But as far as anyone knew, there was not a man alive who had paddled those final 500 miles, via Gods River, clear to Hudson Bay. And before Sevareid and Port reached Norway House and the trackless labyrinth beyond, they had to traverse Lake Winnipeg—275 miles of open water known for the deadly treachery of its sudden squalls, and its wild north winds that could kick up rollers six feet or more in height, swamping a canoe or keeping travelers landbound for days or weeks on end.

Nonetheless, they went, over the initial protestations of parents and teachers. School was behind them, and they had an unshakable determination to accomplish some sort of impossible journey to someplace—to the North Pole or South Africa, or any spot that was hard to get to and far away. It was the sort of dream that most boys are fired with at some stage between youth and manhood, and this was the age of Charles A. Lindbergh, Admiral Richard E. Byrd and the first flights over the North and South Poles. The trip took from mid-June to the end of September, and by the time the two young men got to York Factory they were worrying about freeze-up. The story of their eventful experience is told in *Canoeing with the Cree*, a book Sevareid originally published in 1935.

A few paragraphs from the book illustrate the hardships that men can find in the north woods. It was early September, and the two voyageurs had been paddling for nearly three months: up the Minnesota River southwest, west and finally northwest to the Bois de Sioux, down its reedy channels to the Red River of the North, down the muddy Red for more than 300 miles to Winnipeg, up that fearsome lake to Norway House, the final jumping-off point. Civilization was many days behind them; unknown miles still lay ahead, and for days on end it had been raining. Winter, with all its dangers in the north woods, would soon be setting in.

"Drip . . . drip . . . drip.

"Every time we touched a branch, drops showered upon us. Twenty yards of pushing through the trees and we were drenched again. Now even birch bark refused to burn readily. Dry wood was impossible to find; only by painstakingly cutting out heartwood could we start a fire. We laid our blankets close to the fire, usually with a lean-to above us. Vainly we would try to dry our clothing. By the time it was half fit for wear, our eyes, aching from smoke and weariness, would refuse to stay open. We never failed to awake in pouring rain. Sleep lasted but a few hours each night. The dark hours were dreadfully cold.

"Each day we scanned the skies anxiously, watching for a break in the clouds. But day after day the leaden heavens lowered above. The icy wind blew directly against us time and again. Our faces grew raw and black from exposure. We could not shave, nor even wash our faces. All our wearing apparel was on our backs, day and night. We lost the heels to our rubber-bottom boots and walking over slippery rocks with the canoe over our heads was very dangerous. Once, as we were pulling the canoe along by a rope, Walt slipped and slid backwards down a twelve-foot granite wall. Like a cat, he landed in the center of the boat and by a miracle he did not tip it."

To eke out their dwindling rations, the boys determined to eat only a certain small amount each day. "The animals stayed well away from the river, hidden in their retreats. . . . We saw only a wolf, a black bear and occasionally smaller animals. In an attempt at diversion, we chased a loon two miles along the river one day. One night we sat upright, startled out of sleep by the unearthly scream of a lynx close at hand. We gripped the rifle and waited. But the scream was not repeated, only a low rumbling sound came to our ears and then ceased. Another morning, when we got up, we found the deep impressions of a giant moose in the clay of the river bank. The animal had come to drink that night, only twenty feet from our camp. . . .

"The portages were overgrown. It was evident that no one had gone through the region for many months. Our canoe began to leak badly.

Ragged but erect, three jack pines cling to a small rocky island in one of the myriad lakes in Minnesota's Boundary Waters Canoe Area.

As I was pouring water over the breakfast fire one morning, I saw Walt bending over and peering intently at something in the water of the river. When he beckoned to me, there was a queer look on his face. In a quiet pool, tiny, weblike traceries of shore ice were forming."

The boys' physical misery and the looming threat of winter brought on mental anguish, too.

"Slowly, the unending monotony of the forest gloom, the cold, depressing misty half-light in which we traveled and the unceasing discomforts we underwent, began to fray our nerves. Gradually our dispositions gave way under the strain. We became surly and irritable. The slightest mishap set our nerves jumping. A sudden blast of wind against us, the upsetting of a dish of food, the refusal of wood to burn —these little irritants put us into a rage. Impossible as it appears to us now, we began to vent our ugly moods upon each other. . . . Like children, we bickered."

Under the strain, the two boys actually fell to blows. But then their luck changed. They were still uncertain of their whereabouts, still unsure of how much farther they had to go before ice would render the rivers impassable. They thought they were nearing the Shamattawa, a river whose name means "fast-running water" in Cree and that was virtually the last stage on their journey. Then, suddenly, as they rounded a bend, they saw the remains of a still-smoldering campfire. Tense with expectation, they paddled as swiftly as they could and, as they skirted the next bend, saw a small dot—it had to be a canoe—far ahead on the water. Two hours later, they were attempting to talk sign language to a family of Cree Indians.

They kept repeating, "Shamattawa, Shamattawa," and finally the squaw understood. She gestured toward Sevareid's watch pocket. When he took out the watch, it read 5 o'clock. The squaw took it, pointed to 5:30, and said: "Shmattwa." They were only half an hour away from the fast-running water.

Three days later, their food all gone, they rounded a bend in the Hayes River and saw Hudson Bay spread out before them, with a schooner riding at anchor in the river's mouth. It was September 20. They had been paddling for more than three months and had traversed a wilderness under conditions that most people believed were impossible. Sevareid—a Middle Westerner who had never seen the ocean—remembers thinking, at the moment when they first saw the bay: "This is what all the rivers come to. All those rivers. This is the sea, where everything ends."

Part of Sevareid's and Port's achievement was paralleled at the time this book was being written by two young men from Coon Rapids, Minnesota, who traveled in two kayaks from the Red River of the North in North Dakota to York Factory. Randy Bauer and Gerry Pedersen had had their full share of adventure when I encountered them on the train going south from Churchill, the same line over which Sevareid and Port returned. Their hands were callused and blackened by weeks of paddling and building campfires: the trip had taken them 61 days. And just a few days before they arrived at York Factory a party of six, in three canoes, had also come in: they too had begun their journey at the Red River of the North.

Bauer and Pedersen kept a journal of their voyage. But the words of a diary written by the light of candles or a campfire are sparse; they can sketch in only the outline of the days and nights spent in a huge wilderness, with stars singing overhead or rain pelting down, and with only a thin-skinned tent for protection against the cold and damp. Out there in the woods the world is an enormous place. Yet the sentiments felt by those who make such journeys are always the same. "Kind and harsh this northland is," wrote Randy Bauer one night in his tent on Windy Lake. "We curse it and love it at the same time." And in 1938 Florence Page Jaques, wife of the north-woods artist Francis Lee Jaques, wrote of a campsite she had come to on her first canoe trip: "It's not our world at all; it's another star."

This sense of being removed, taken right out of one's ordinary life and being transported to a different world, is part and parcel of canoe travel in the north woods. Much of the feeling stems, I think, from the abruptness of the transition: one steps from car or train or airplane into this fragile shell of canvas, aluminum or fiberglass, paddles to the first portage, beyond which the outboard-motor boats cannot go, and suddenly the familiar world is gone, and there is nothing but the primitive and the wild all around. There is no way to familiarize oneself gradually with this wilderness; one is simply dropped into it, and there is no continuity of history to soften its hard edges. As Mrs. Jaques wrote: "This country never knew a medieval time; it came straight from the primeval into today."

It is not necessary to travel into the far northern reaches of Canada to find this solitude. It lies everywhere in the canoe country, and the marvel is how little effort is required to enter that world and achieve a sense of harmony with it. I remember a trip I took once, with my brother and my 14-year-old son, into the canoe country along the Canadian border north of Gunflint Lake. There we found an island of genuine enchantment, an island that, as far as we could tell on arrival, had never before been trod by human feet.

We started our trip one August afternoon from a point south of Saganaga Lake, planning a journey of about a week that would take us in a large circle through Saganaga and several other lakes back to our starting point. There were a few cars parked there when we left, indications that other canoe campers had departed into the wilderness before us. Yet in our entire trip we saw another human being only once, and that was from a distance.

Ours was an aluminum canoe, and with two tents, a jungle hammock, sleeping bags, cooking equipment and food for three it was comfortably loaded. We pushed off into a small stream that led through a tiny set of rapids past a point and into Saganaga Lake itself. Within 15 minutes we had left the world we knew.

Overhead a raven croaked, and far above him a hawk sailed in endless circles, rising on a column of warm air. The water was purest blue and crinkled with millions of tiny ripples stirred by a vagrant breeze. Through the days that followed, I paddled in an utterly dreamlike state. My mind revolved around food, campsites and the lake, from which we drank the purest of water whenever we felt thirsty, simply by cupping our hands and dipping it up as we drifted along in the canoe. By 5 or 6 o'clock every afternoon our campsite would appear; we never knew where it might be, but we knew that we would find a place before darkness set in. We landed, pitched our tents, built our fire, cooked our supper, and then we sat around and smoked and talked until, half asleep, we crawled off to bed. And traveling leisurely along in this way we came to the special island.

It was a typical product of the Canadian Shield, this island that was to become uniquely ours: a rounded, shelving, rocky ledge at one end, a clump of pines and soft, mossy ground at the other. There are a thousand variations of this kind of island all through the lake country, shaped eons ago by glacial action, but we were soon to find certain important differences.

Its rocky end was not only nicely shelved, affording an excellent unloading site, but it also had a troughlike indentation at the water's edge that could have been especially made for a canoe. It was like coming into a slip: we poked our bow in, it came to rest on a rocky incline, and

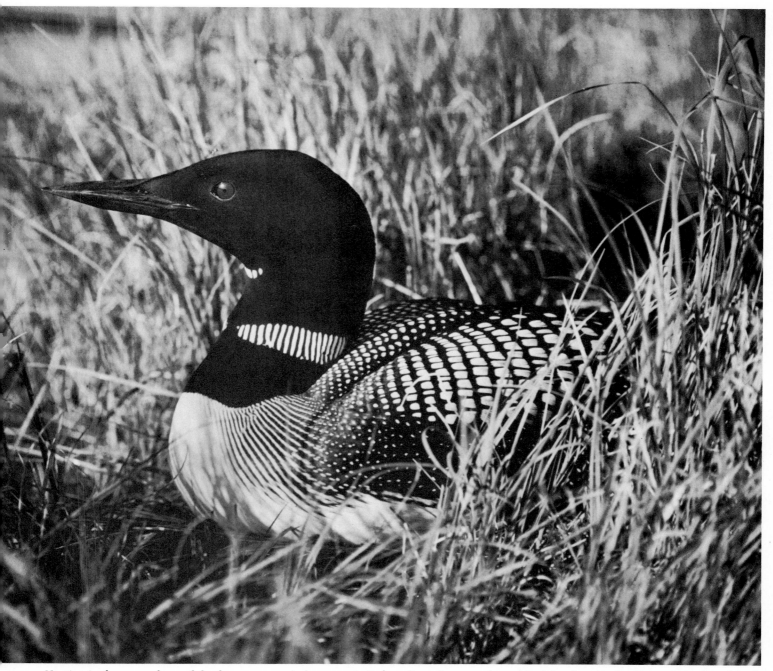

Nesting in the grass along a lakeshore, a common loon sports its breeding plumage: glossy black head, striped collar and checkered back.

we stepped ashore without so much as wetting the soles of our sneakers. We could draw the canoe up safely out of reach of even the stormiest waves. Loading and unloading were as simple as if the canoe were lying alongside a dock, and there was never any fear of carelessly capsizing her as we put our gear ashore.

Just beyond the canoe landing, a little closer to a clump of trees, was another convenience: a level ledge right next to the water that was a perfect place for washing dishes. This is a boon not to be taken lightly, for dishwashing is a chore that must be performed right after a meal, and finding a place where one can comfortably wash the dishes and then stack them without having them slide or roll off into the water is often difficult. Never, in all my subsequent experience, did I find a dishwashing site that came close to that one.

The living quarters on the wooded end of the island were superb. There was level ground at the highest spot where our two small tents and the jungle hammock could be set up secure against wind, and the ground there was springy and soft with brown pine needles. There was no fear of having rain water invade the tent in a storm because the ground sloped off on all sides. A dead tree had fallen precisely where we needed it to sit on; in front of it we built our campfire on a bed of stones we found conveniently at hand.

Finally, we discovered a perfect place for swimming, the best we had ever seen. Across the ledge from the canoe landing, the rock fell steeply down into the lake. The top of this little cliff ranged from five to 10 feet above the water line; it was a natural place on which to stand. And there was no problem of falling on slippery rock when climbing out: the glacier that passed by long ago had obligingly plucked chunks off the rock, leaving a series of rough steps—a natural staircase on which we could comfortably return to land.

What does a person do in a place like that perfect island to pass the time? Actually, the question never arose. I cannot remember a moment in which I was bored, and neither can my brother or son. For that matter, I cannot remember a moment when I measured time. The first couple of days, I dutifully wound my watch every night. On the third night, I thought to myself: "Why wind it? Why not give it a rest? Why not try to see what it's like to live without measuring time?" And for the balance of the trip that was exactly what we did—we lived like deer or loons, measuring our days by the sun, eating when we were hungry, falling asleep when we were tired, waking up when we had slept long enough. There is no better restorative for a mind and body wearied by the pace of civilized living.

The first thing, every morning, was a plunge in the lake. The lake took care of just about everything that had to do with bodily cleanliness: I washed in it, I brushed my teeth in it, I shaved in it. The water was so clear that we could see the bottom 20 feet below from the cliff on which we stood to dive. There were rocks down there, and the sunlight played on them in beautiful effects of light and shadow filtered through the wavering blue. The water seemed almost a continuation of the air in liquid form; swimming in it was like jumping off a mountainside and gliding, like a gull.

Day after day dawned beautiful and sunny. We would emerge from the lake, shaking the water off our skins like dogs, and build the fire for breakfast. From then until nightfall, the fire never went out; it might sink down to glowing embers, but someone would always put a few branches on in time to keep it going. In the evening, from my sleeping bag, I would see its warm red coals glowing through the entrance of my tent, and I always went to sleep with this vision of man's first primitive warmth and comfort.

What did we eat? One morning we ate three fish that I caught before breakfast—three lovely smallmouth bass, about a pound each. We baked corn muffins, we made pancakes. My brother made bannock, the unleavened bread that has for centuries been a staple of the north woods; it was a skill he had acquired on previous trips. I don't remember the menu that we put together from our stores; I remember only that it was, quite literally, the best food I ever ate because I was hungry for it every day.

We had endless things to do. We fussed with our tents until they were absolutely perfect; we could have lived in them for years. We gathered wood and pine cones for the fire, stacking it neatly, with the cones at one side in a little pile. We washed our clothes and hung them up to dry. When we were sleepy, we took naps. When we woke up, we dived into the lake again.

And one day, down among the rocks on the wooded end of the island, we found a broken paddle. It was a little like Robinson Crusoe's finding Friday's footprint on the sand: suddenly we realized that we were not alone in the world. Where the paddle came from we had not the slightest idea; it could have been an accident that happened 50 weeks before, or 50 years. The paddle was old and silver gray from exposure. Totally by accident, we had found a name for our island: Broken

Paddle Island. We carved the name in the paddle, added our own names and leaned it against a tree.

I have been on many campsites since that magic time on Broken Paddle Island, and I have seen the other side of the coin. Sometimes there simply is no decent place to set up a tent and cook dinner, and on such occasions I have learned to make the best of what there is. There are lakes with marshy banks from which mosquitoes rise in dense stinging clouds as soon as one sets foot ashore; such places are to be tolerated only as long as absolutely necessary, because there is no use trying to fight the insects of the north woods. Get into bed and fall asleep as fast as possible; rise early and flee—that is the only maxim to follow in those situations. The bad nights must be taken with the good, and what one campsite lacks the next will probably have; this is, for me, the best camping country in the world.

Nor should an occasional day of rain spoil the pleasure of a canoe trip. The world takes on a unique beauty when rain is falling on the lakes and forests. Colors are deeper and very pure. The mosses, saturated with water, are rich and full; even the lichens turn thick and spongy and assert their character as living plants rather than just scabs on the rocks. The trees glisten in the soft light, each leaf and needle a resting place for glittering drops of water. The light diffuses softly through and over everything—there are no contrasts, no shadows, only an abiding softness in the shapes and outlines of the landscape that matches the pervading silence. The light reflects dimly off wet roots twisting in the earth, and from wet rocks at the water's edge. It is no hardship to paddle a canoe under these conditions, and the woods may show things that otherwise one might never see.

For example: fog blowing off the hills and through the treetops as though the land were breathing. A distant grove of aspens glowing a pale green among the dark green pines. A duck speeding low over the lake, wings beating a mad tattoo in the quiet air. A deer sticking its nose out of the woods, then advancing to the water unafraid as the canoe glides by.

The water on such a day is the color of beaten silver. I remember a rainy day on the Granite River when, paddling from one campsite to another, I stopped to study the water as carefully, as reverently, as I might study a work of art. The sound it made as it slipped down among the rocks was a soft gurgling, a quiet talking that reminded me of a child talking to itself as it sits engrossed in play, or the sweet mur-

Mature red and white pines—the reds are oval-crowned and shorter than the whites—dominate the rocky shoreline of Lac La Croix.

murings of lovers. It was a spiritual talking, expressed in tones of marvelous liquidity as the river slowly dimpled its way down through rounded rocks toward the rapids below.

Here the talk became more boisterous. There was conflict now between the water and the rocks, and the river dipped to meet the challenge. Now it rose in a swirl as it met the big rocks, and its talk changed to a continuous protesting and arguing. Waves curled up from the smooth surface, then turned white and foamy as they broke in the rapids. Then, suddenly, as we coursed through this small section of rapids, the water was still again.

It was here on this gray and quiet day that I fished a drowning bee out of the water. I saw it struggling feebly, and as we approached I slid my paddle gently under it and lifted it into the silvery air. The bee was a pitiful sight; it lay there motionless, seemingly unaware that it was out of danger. I held the paddle close to my face to observe it, and I could see the fat black-and-yellow thorax pulsing, exactly as if it had been a spent animal fighting for breath after a hard chase. It lay there for a considerable length of time—and then, suddenly, with what was obviously an enormous effort, it staggered to its feet and stood there shakily. Its body heaved, its legs trembled. I could see its wings, all sodden with water, lying flat along its belly. For the time being, it made no effort to raise them.

By this time I was utterly absorbed in watching the struggle for life of this tiny creature. It seemed so courageous, standing there trembling, but still trying to regain control of its water-soaked body. The next move came from its antennae. There was a sudden twitch, then slowly, creakily, the drooping black threads came to life, stiffened, and moved up and forward. The black antennae turned this way and that, probing the air. Now there was a sudden blur of movement along the bee's belly: the first attempt to use its wings. The effort lasted only a split second; those delicate gossamer members must have felt to the exhausted bee as though they were made of lead. But now the bee raised its yellow hind legs, one after the other, and set about cleaning its wings, stroking them, straightening them, shaking tiny droplets of water off them. And once again it tried to use them—but it was too soon. A blur of movement, and they were still again.

Now the bee began to crawl along the paddle. Its agility was quite surprising. It seemed to be trying to find out where it was. It crawled to the very edge of the paddle, then stopped and seemed to lean out over it. Below, the water was slipping past, smooth, silver and treacherous. Did the bee see it, and remember?

Again it rested, for perhaps half a minute. Then it tried its wings once more. They were still too wet, and so it set about again to clean them. Back and forth the hind legs worked, stroking, smoothing, shaking off the droplets that still held its wings captive. Now it took a few more shaky steps. Now the wings blurred into life once more, and then again. And now the bee rose slowly into the air. For a moment it hovered over the paddle, the wings a blur of silver against the black-and-yellow body; then, with a sudden forward dart, it was gone across the water, back toward the shore.

Such are the rewards of the gray days, when in the soft, damp stillness one can draw close to the essence of the forest and its inhabitants. These are the times of intimate looks, when the forest world seems open and receptive, when its busy work is slowed and strangers may come in. This is a time of somber beauty in the woods, when the present fades away into the silence.

There are other days, too. I stood one autumn afternoon in a yellow-golden meadow in Manitoba, not far from Sioux Lookout. The air this day was very still, with only the faintest of wandering breezes ruffling the grass and moaning gently in the treetops. The place smelled heavily of mushrooms, damp earth, rotting leaves and pine needles. Birds chirped faintly now and then, but their season was passing and they were few in number. The woods were dark and cold; it was as though here, in this meadow, the essence of summer was still lingering—one last look, one last warm breath before the winter finally closed down.

The Forest's Last Stand

Where the north woods end and the treeless tundra begins there is no sharp line of demarcation. The two environments intermingle in a subarctic zone up to 200 miles wide, and the balance between them shifts gradually, so that a northbound traveler does not get the full impact of the change until he reaches the zone's ragged upper edge. The forest's dramatic last stand is shown here in pictures taken near Churchill, a town on the western shore of Hudson Bay founded by 17th Century English fur traders.

In this bleak landscape, spruces are no longer the giants of the densely wooded terrain to the south; under the savage scourging of the subarctic climate they dwindle to small, scattered clumps. Some trees stand stunted and misshapen in sheltered pockets of low-lying muskeg or bog amid stretches of naked or lichen-crusted rock. Other trees, exposed to the powerful prevailing wind, grow aslant with branches only on their leeward side.

The earth itself seems to be working against the trees. Underlaid by permafrost and insulated against the summer sun by a thick mat of tiny plants, the ground thaws out to a depth of only three to five feet. In this thin layer of poor soggy soil, trees must spread their roots wide both to stay upright and to absorb meager nutriment. Expectably, the trees that dominate this terrain are shallow-rooted, water-tolerant species, chiefly black spruce and tamarack. White spruce, aspen and alder grow on better-drained sites.

The changing of the seasons does little to relieve the trees' ordeal. Spring is ephemeral, and in summer the trees are drenched not only in meltwater but also in 10 inches of rain. September usually brings the first snowfall. Soon afterward, the Churchill area is once more seized in winter's merciless grip, not to be released for almost eight months.

It would be a triumph if the trees merely held their own in this hostile world. But they may be doing better than just that. Scientists calculate that the ice age glaciers left the Hudson Bay region stripped bare only 7,300 years ago. They consider it a trend that, with the climate warming up in the relatively short span of time since, the plant succession has already created new soil and resettled the area. Thus the mere presence of trees this far north suggests that the forest may still be advancing, invading the tundra's domain.

MARCH: BUFFETED SPRUCE BRANCHES

MARCH: A LONE WHITE SPRUCE FRAMED BY SNOW RIDGES

JUNE: AN ALDER REFLECTED IN MELTWATER

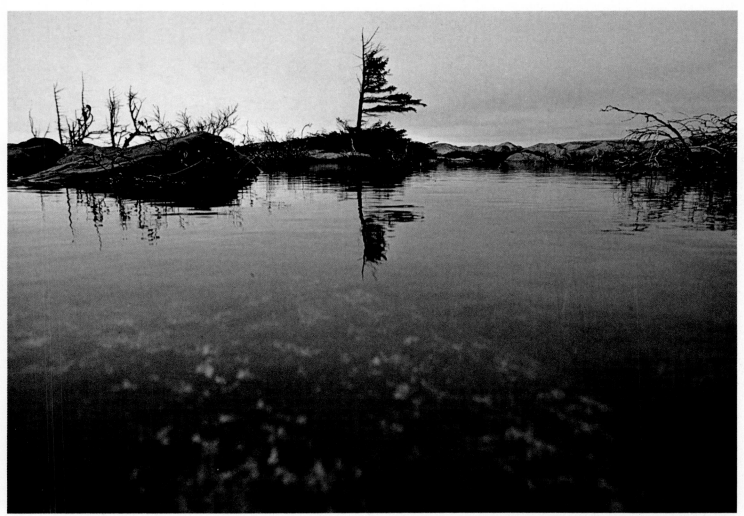

JUNE: A WHITE SPRUCE PIONEERING A ROCKY OUTCROP

JUNE: ROCK AND ICE ALONG HUDSON BAY

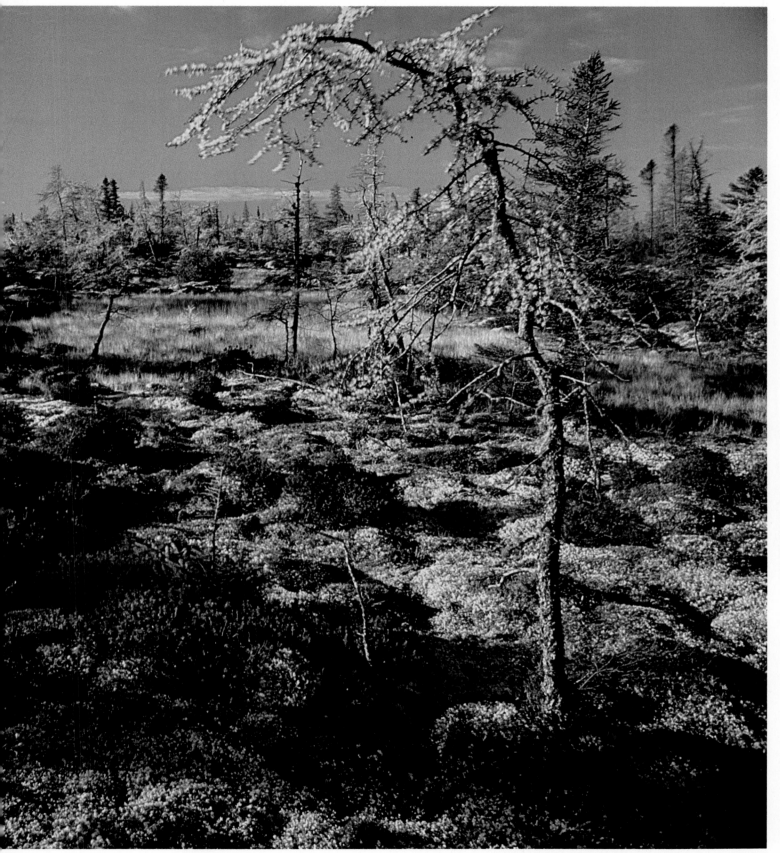

OCTOBER: BLACK SPRUCES, TAMARACKS AND LICHENS

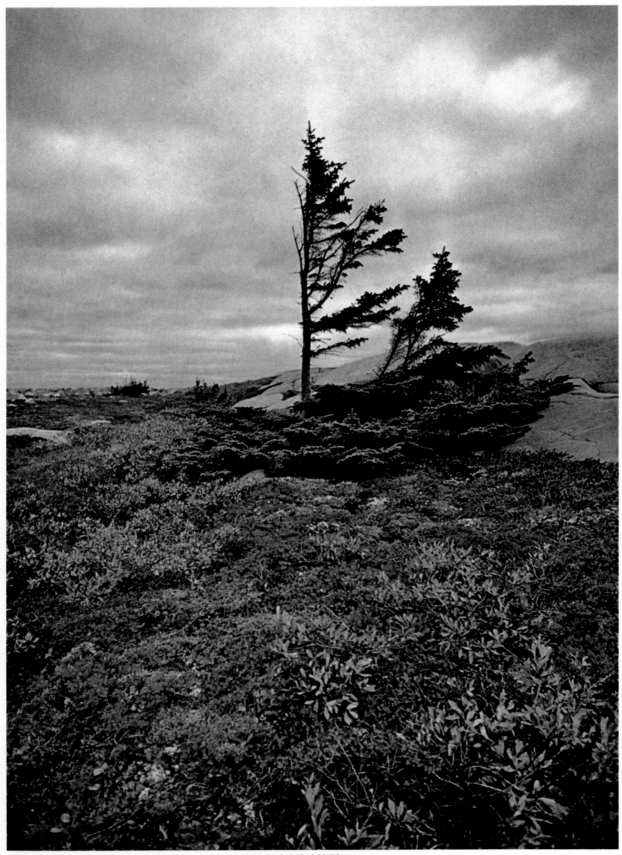

OCTOBER: A SPINDLY WHITE SPRUCE AND SUBARCTIC GROUND COVER

OCTOBER: MUSKEG UNDER A FOREBODING SKY

MARCH: WHITE SPRUCES BURDENED BY SNOW

MARCH: A SNOW FORMATION AT THE BAY'S EDGE

Bibliography

*Also available in paperback
†Only available in paperback

Bland, John H., *Forests of Lilliput: The Realm of Mosses and Lichens.* Prentice-Hall, Inc., 1971.

Bolz, J. Arnold, *Portage into the Past.* University of Minnesota Press, 1960.

Burt, William H., and Richard P. Grossenheider, *A Field Guide to the Mammals.* Houghton Mifflin Company, 1964.

Cahalane, Victor H., *Mammals of North America.* The Macmillan Company, 1947.

Cobb, Boughton, *A Field Guide to the Ferns.* Houghton Mifflin Company, 1956.

Collingwood, G. H., and Warren D. Brush, *Knowing Your Trees.* The American Forestry Association, 1964.

Dewdney, Selwyn, and Kenneth E. Kidd, *Indian Rock Paintings of the Great Lakes.* University of Toronto Press, 1967.

†Douglas, William O., *My Wilderness: East to Katahdin.* Pyramid Books, 1968.

Farb, Peter, and the Editors of TIME-LIFE BOOKS, *The Forest.* TIME-LIFE BOOKS, 1963.

Farb, Peter, and the Editors of TIME-LIFE BOOKS, *The Land and Wildlife of North America.* TIME-LIFE BOOKS, 1966.

*Hale, Mason E., *How to Know the Lichens.* Wm. C. Brown Company Publishers, 1969.

*Innis, Harold A., *The Fur Trade in Canada.* University of Toronto Press, 1962.

Jaques, Florence Page, *Canoe Country.* University of Minnesota Press, 1938.

Kimble, George H. T., and Dorothy Good, eds., *Geography of the Northlands.* The American Geographical Society, 1955.

Lakela, Olga, *A Flora of Northeastern Minnesota.* University of Minnesota Press, 1965.

Lawrence, R. D., *Wildlife in Canada.* Michael Joseph Limited, 1966.

MacKay, Douglas, *The Honourable Company.* The Bobbs-Merrill Company, 1936.

MacLennan, Hugh, *The Rivers of Canada.* Charles Scribner's Sons, 1961.

Mech, L. David, *The Wolf: The Ecology and Behavior of an Endangered Species.* Natural History Press, 1970.

Meen, V. B., *Quetico Geology.* University of Toronto Press, 1959.

Moon, Barbara, *The Canadian Shield.* N.S.L. Natural Science of Canada Limited, 1970.

†Morse, Eric W., *Canoe Routes of the Voyageurs.* Minnesota Historical Society, 1962.

Morse, Eric W., *Fur Trade Canoe Routes of Canada/Then and Now.* Minnesota Historical Society, 1969.

Murie, Olaus J., *A Field Guide to Animal Tracks.* Houghton Mifflin Company, 1954.

Nelson, Bruce, *Land of the Dacotahs.* University of Minnesota Press, 1946.

†Nute, Grace Lee, *Rainy River Country.* Minnesota Historical Society, 1950.

Nute, Grace Lee, *The Voyageur.* Minnesota Historical Society, 1955.

Nute, Grace Lee, *The Voyageur's Highway.* Minnesota Historical Society, 1965.

Olson, Sigurd F., *Listening Point.* Alfred A. Knopf, 1970.

Olson, Sigurd F., *The Lonely Land.* Alfred A. Knopf, 1961.

Olson, Sigurd F., *The Singing Wilderness.* Alfred A. Knopf, 1957.

Olson, Sigurd F., and Les Blacklock, *The Hidden Forest.* Viking Press, 1969.

O'Meara, Walter, *The Savage Country.* Houghton Mifflin Company, 1960.

Orr, Robert T., *Mammals of North America.* Doubleday and Company, 1971.

Petrides, George A., *A Field Guide to Trees and Shrubs.* Houghton Mifflin Company, 1958.

Pruitt, William O., Jr., *Animals of the North.* Harper and Row, 1960.

*Robbins, Chandler S., Bertel Bruun and Herbert S. Zim, *Birds of North America.* Golden Press, 1966.

Rue, Leonard Lee III, *The World of the Beaver.* J. B. Lippincott Company, 1964.

Rutter, Russell J., and Douglas H. Pimlott, *The World of the Wolf.* J. B. Lippincott Company, 1968.

Seton, Ernest Thompson, *Life Histories of Northern Animals,* Vol. I. Charles Scribner's Sons, 1909.

Sevareid, Eric, *Canoeing with the Cree.* Minnesota Historical Society, 1968.

Silvics of the Forest Trees of the United States. U.S. Department of Agriculture, Forest Service, 1965.

Smith, Alexander H., *The Mushroom Hunter's Field Guide.* University of Michigan Press, 1963.

Spears, Borden, ed., *Wilderness Canada.* Clarke, Irwin and Company, Limited, 1970.

Steinhacker, Charles, *Superior: Portrait of a Living Lake.* Harper and Row, 1970.

Taverner, P. A., *Birds of Canada.* National Museum of Canada, Canada Department of Mines, 1934.

Warren, Edward Royal, *The Beaver, Its Work and Its Ways.* The Williams and Wilkins Company, 1927.

Wilsson, Lars, *My Beaver Colony.* Doubleday and Company, 1964.

Acknowledgments

The author and editors of this book are particularly indebted to William O. Pruitt Jr., Professor of Zoology, University of Manitoba, Winnipeg. They also wish to thank: Donald F. Bruning, Assistant Curator of Ornithology, New York Zoological Park, New York City; Selwyn Dewdney, London, Ontario; Harold S. Feinberg, Department of Living Invertebrates, The American Museum of Natural History, New York City; Dorothy Gimmestad, Minnesota Historical Society, St. Paul; Miron L. Heinselman, Principal Plant Ecologist, North Central Forest Experiment Station, U.S. Forest Service, Ely, Minnesota; Douglas Hornby, Director, Peace-Athabasca Delta Project, Edmonton, Alberta; Sidney S. Horenstein, Department of Invertebrate Paleontology, The American Museum of Natural History, New York City; John W. Miller, New York City; Eric W. Morse, Tenaga, Quebec; Charles W. Myers, Department of Herpetology, The American Museum of Natural History, New York City; John Pallister, Department of Entomology, The American Museum of Natural History, New York City; Larry Pardue, The New York Botanical Garden, New York City; Douglas Pimlott, Professor of Animal Ecology and Wildlife Biology, University of Toronto, Ontario; Russell J. Rutter, Huntsville, Ontario; Alexander Smith, Professor of Botany and Director, University Herbarium, University of Michigan, Ann Arbor; Frederick Ward, Professor of Zoology, University of Manitoba, Winnipeg.

The translation of the song "Quand un Chrétien Se Détermine à Voyager," pages 72-73, is from *The Voyageur*, pages 151-154, by Grace Lee Nute, Minnesota Historical Society, 1955 (© 1931 D. Appleton & Co.).

Quotations, pages 152, 154, are from *Canoeing with the Cree*, pages 161-166, by Eric Sevareid, Minnesota Historical Society, 1968 (© 1968 Eric Sevareid).

Picture Credits

Sources for pictures in this book are shown below. Credits for pictures from left to right are separated by commas, from top to bottom by dashes.

Cover—Gerald R. Brimacombe. Front endpapers 2, 3—Gerald R. Brimacombe. Front endpaper 4, page 1—Jim Brandenburg. 2, 3—Paul Jensen. 4, 5—Les Blacklock. 6, 7—Harald Sund. 8 through 11—Charles Steinhacker. 12, 13—Paul Jensen. 18, 19—Map by R. R. Donnelley Cartographic Services. 24—Paul Jensen. 29—Barbara K. Deans, John Kohout from Root Resources—Dr. E. R. Degginger. 30—Larry West—Dr. W. A. Crich—G. J. Harris, Dr. E. R. Degginger. 31—Dr. E. R. Degginger—Barbara K. Deans, Dr. W. A. Crich (2). 37—*The Minneapolis Tribune*. 38 through 41—Gerald R. Brimacombe. 42, 43—Gerald R. Brimacombe, Les Blacklock. 44—Paul Jensen. 45, 46, 47—Les Blacklock. 50, 51—G. J. Harris. 54—Gerald R. Brimacombe. 56, 57—Gerald R. Brimacombe. 59—Bruce Litteljohn. 64—from *Harper's Weekly*, Feb. 1, 1890. 68, 69—Jim Brandenburg. 75—Paulus Leeser courtesy Public Archives of Canada. 76, 77—Glenbow-Alberta Institute. 78 through 81—Paulus Leeser courtesy Public Archives of Canada. 84—T. W. Hall. 87—Les Blacklock. 90—Jim Brandenburg. 91—Lynn L. Rogers. 95—David Cavagnaro. 96, 97—Kenneth W. Fink, David Cavagnaro—Craig Blacklock. 98 through 105—David Cavagnaro. 106, 107—Kenneth W. Fink. 112, 113—Les Blacklock. 117—D. Mohrhardt from the National Audubon Society—W. Victor Crich, Jim Brandenburg (2). 121, 122, 123—Harald Sund. 124, 125—Ed Cesar from The National Audubon Society. 126—Les Blacklock. 127—Les Blacklock (2)—L. David Mech, Gilbert F. Staender. 128, 129—Harald Sund. 134, 135—Robert Walch. 136—Les Blacklock. 140 through 149—Robert Walch. 152, 153—Jim Brandenburg. 157—James H. Bellingham. 160, 161—Gerald R. Brimacombe. 165, 166, 167—Harald Sund. 168, 169—Gerald R. Brimacombe. 170, 171—Norman R. Lightfoot. 172 through 175—Paul Jensen. 176 through 179—Harald Sund.

Index

Numerals in italics indicate a photograph or drawing of the subject mentioned.